Frederick Overman

The Moulder's and Founder's Pocket Guide

A treatise on moulding and founding - With an appendix containing receipts for alloys, bronze, varnishes, and colours for castings

Frederick Overman

The Moulder's and Founder's Pocket Guide
A treatise on moulding and founding - With an appendix containing receipts for alloys, bronze, varnishes, and colours for castings

ISBN/EAN: 9783337139360

Printed in Europe, USA, Canada, Australia, Japan

Cover: Foto ©Andreas Hilbeck / pixelio.de

More available books at **www.hansebooks.com**

THE

MOULDER'S AND FOUNDER'S POCKET GUIDE:

A TREATISE ON

MOULDING AND FOUNDING IN GREEN-SAND, DRY-SAND, LOAM, AND CEMENT; THE MOULDING OF MACHINE FRAMES, MILL-GEAR, HOLLOW-WARE, ORNAMENTS, TRINKETS, BELLS AND STATUES; DESCRIPTION OF MOULDS FOR IRON, BRONZE, BRASS, AND OTHER METALS; PLASTER OF PARIS, SULPHUR, WAX, AND OTHER ARTICLES COMMONLY USED IN CASTING; THE CONSTRUCTION OF MELTING FURNACES, THE MELTING AND FOUNDING OF METALS; THE COMPOSITION OF ALLOYS AND THEIR NATURE.

WITH AN APPENDIX

CONTAINING RECEIPTS FOR ALLOYS, BRONZE, VARNISHES AND COLOURS FOR CASTINGS, ALSO TABLES ON THE STRENGTH AND OTHER QUALITIES OF CAST METALS.

BY FRED. OVERMAN,
MINING ENGINEER.

AUTHOR OF 'THE MANUFACTURE OF IRON,' "A TREATISE ON STEEL,' ETC. ETC.

With Forty-two Wood Engravings.

PHILADELPHIA:
HENRY CAREY BAIRD & CO.,
INDUSTRIAL PUBLISHERS,
810 WALNUT STREET.
1878.

Entered, according to the Act of Congress, in the year 1851, by
A. HART,
in the Clerk's Office of the District Court of the United States, in and for the Eastern District of Pennsylvania.

CONTENTS.

CHAPTER I.

MOULDING.

Materials for Moulding.—Sand, 14; green-sand, 14; dry sand, 15; core-sand, 16; clay, 17; loam, 18; blackening, coal-dust, black-lead, anthracite, 19; soapstone powder, 20; localities of materials, 21; grind-mills for blackening, 22.

Tools.—Flasks or boxes, 23; crane, 25; small tools, trowels, cleaner, stamper, shovels, sieves, 28; pease-meal, parting-sand, gate-pins, screws, &c., 30.

Moulding in Green-Sand.—Moulding of a wheel, 31; filling in the drag-box, 32; the top box, 36; gits, 36; removing the top box, 38; drawing of the pattern, 39; blackening the mould, 41; pattern fastened to the moulding-board, 43; composition of moulding-sand, 43; general observations, 43; division of labour, 43.

Moulding in Open Sand.—Making of the bed, 48; moulding in one box, 50; moulding of a cog-wheel, 52; causes of failures, 57; moulding in more than two boxes, 59; small articles of machinery, 61; ornamental moulding, 62; hollow-ware, 66; small articles, 67; moulding of a coffee-kettle, 68; patterns for hollow-ware, 70.

Mixed Moulding.—Moulding in green-sand with dried cores, 73; cores and their use, 74; core-boxes, 75; moulding of a column, 76; making a pipe-core, 80; moulding with plates, 82.

Dry-Sand Moulding.—Drying of the mould, 84; glands, 85; vertical castings, 86; moulding of a large pipe, 87; loam core, 88; hay ropo, 88; core-iron, 89; making of the core, 89; casting pipes without a core, 93; small ornamental castings, 94; trinkets, 94; moulding a stag, 96; screwing together of ornaments, 99; soldering, 99; brass castings, 99; fine iron castings, 99.

Loam-Moulding.—Quality of loam, 101; compounding loam, 101; moulding of simple round forms, 104. *Moulding of a soap-pan,* 104; the core, 105; the thickness, 108; the cope, 108; taking apart the mould, 110; blackwash, 110; gates, 112; gas-pipe, 112; cast-gate, 112; covering of the flow-gate, 113; use of the flow-gate, 114; casting by a single gate, 115; gas-pipes, 115.

Moulding without Thickness.—A steam cylinder, 116; making of the cope, 117; the steam-ways, 117; cores for the steam-ways, 121; core for the cylinder, 122; sullage piece, 123; fastening of the cores, 124; burying of the mould, 125; fastening of the exhaust-pipe core, 126; use of chaplets, 126; general remarks on cylinder moulding, 127.

Irregular Forms.—Moulding of a curved pipe, 128; oval forms, 132; bathing-tub, 132; elbow-pipe, 133; complicated forms, 134.

Moulding of Bronze Ornaments.—Moulding of statues by the ancient Greeks, 138; French mode of moulding statues, 139; present mode of casting statues, 141; iron statues, 142; bas-reliefs, 142; moulding of bells, 143.

Metal Moulds.—The bore of wheels, 146; railroad-car wheels, 146; chilled rollers, 148; casting together of iron and steel, 152; moulds for tin, lead, pewter, zinc, types, &c., 153; mould for copper or brass, 154; plaster of Paris moulds, 155; stereotyping, 155.

Impressions and Castings.—Wax, crumbs of bread, sealing-wax, 159; sulphur, 160; glass, 161; clay, artificial wood, 162; plaster of Paris, 163; mould of a coin, 165; mould of a statue, 168; casting of plaster, 172; taking of a mask, 173; sulphur castings, 174; wax castings, sealing-wax, and other casts, 175; elastic mould, 175; alum, saltpetre, moulding of natural objects, 176.

CONTENTS.

CHAPTER II.
MELTING OF METALS.

Iron.—Qualities of iron: No. 1 iron, 179; No. 2 iron, No. 3 iron, 181; characteristics of foundry pig, 181; mixing of iron, 184; kind of castings, 187; kind of moulds, 188; melting of iron, 188; in the blast furnace, 188; in crucibles, 192; in reverberatories, 196; the cupola, description of, 201; operation in a cupola, 205; pots, 210; blast-machines, 212; fans, 212; hot blast, 217; drying stoves, 217.

General Remarks.—Cleansing of castings, 218; time of casting, 220, cost of moulding and casting, 221; melting of bronze metal, 223; melting of lead, tin, antimony, and brass, 224.

APPENDIX—RECEIPTS AND TABLES.

Alloys of Iron.—Iron and sulphur, iron and carbon, iron and phosphorus, 226; iron and silicon, iron and arsenic, iron and chromium, iron and gold, iron and silver, iron and copper, iron and tin, iron and lead, 227.

Alloys of Precious Metals, 228.

Alloys of Copper.—Bronze, 228; bell metal, 229; bronze for guns, 229; bronze for statues, 230; bronze of the ancient Greeks, 231; bronze of the Aztecs, speculum metal, speculum metal of Rosse's telescopes, 232; bronze for medals, bronze imitation of gold, 233. *Brass*—Common brass, solder, button-brass, red brass, princes' metal, brass and lead, tempering brass, 234; brass for ship nails, brass for pans and steps, brass and platinum. German silver, Chinese packfong, argentan for plating, 235; electron, solder for German silver, copper and platinum, copper and silver, copper and antimony, 236; copper and carbon, copper and phosphorus, copper and arsenic, 237.

Lead and its Alloys.—Lead and arsenic, lead and antimony, stereotype metal, fusible metal, 238.

Tin and its Alloys.—Tin and lead, pewter, queen's metal, Britannia metal, German tin, music metal, antifriction metal, spurious silver organ-pipes, 239; imitation of diamonds, tin foil, 240.

Zinc and its Alloys, 241.

Bronzing.—Natural bronze, antique bronze, 241; various colours, 242; bronze paint, 243.

Gilding of bronze and brass, 243; of iron, 244.

Tinning of brass, bronze, and copper, 245.

Zinking of copper or bronze, 245.

Glazing of castings, enamel, 246.

Blackening of iron with plumbago, 247; **with varnish, 247.**

Grinding and polishing, 248.

Malleable iron, 248.

Silvering of iron, 249.

Table I. 250.
    II. 251.
    III. 251.
    IV. 252.

THE MOULDER'S AND FOUNDER'S POCKET GUIDE.

CHAPTER I.

MOULDING.

The moulding of metals and other materials into the various forms, required for the accomplishment of certain purposes—whether of an economical or ornamental character—is an object of high interest. Moulding is the noblest of the arts; it serves with unvaried interest the fine as well as the useful arts. The heavy castings for the construction of machinery, to the weight of thirty tons and more; the statues of the ancients, and of modern heroes, are ornaments of human genius. The minute, well finished castings of iron and bronze are evidences of human skill and ingenuity.

Moulding may be considered in two distinct branches; the one is the moulding proper, the other the forming of the pattern. Moulding proper is almost the same in principle and in practice for each of the various kinds of metals or alloys. Slight variations in the materials for moulding, and in treatment, are the only differences in moulds which are designed to be used for iron, brass, bronze, tin, or lead, and other metals. The principal materials used in moulding, are, sand of various kinds, loam, plaster of paris, blackening, and metal.

Sand is the most common, and certainly the most perfect and convenient material. It is superior to all other materials for moulding. Sand is more or less porous, and very refractory, so that the hot metals do not melt nor bake it; two qualities of great importance in the successful operations of the business. The various kinds of good moulding sand, employed in foundries for casting iron or brass, have been found to be of an almost uniform chemical composition, varying in grain or the aggregate form only. It contains between 93 and 96 parts of silex or grains of sand, and from 3 to 6 parts of clay, and a little oxide of iron, in each 100 parts. Moulding sand which contains lime, magnesia, and other oxides of metal, is not applicable, particularly for the casting of iron or bronze. Such sand is generally too weak or too close; it will not stand, or retain its form, or it will cause the metal to boil by its closeness. In practice the different classes of castings require different kinds of sand for the purpose of moulding. For one kind of castings the sand is to be porous, open, and is still to be adhesive; for another class it is to be very adhesive and fine, almost free of grit, to make itself conform to the minutest parts of the pattern imbedded in it. At the proper places in the description of the process of moulding, we shall allude to the various kinds of sand best qualified for specific purposes.

The best moulding sand is generally found along the banks of large rivers; that procured from the shores of mountain streams, is in most cases too coarse or too poor and weak. Good sand, however, has been found on the very top of high hills. The best is generally found in the vicinity of the primary rocks, or along those river banks which receive their supply from the primitive mountains. The alluvium of the transition or metamorphous rocks, as graywacke, slate, and feldspar, forms a very superior moulding sand, if it does not contain too much iron. In the coal districts there is generally little or no difficulty in finding good sand, for most of the river flats are composed of that useful material, which, however, frequently contains too much iron, and is liable to melt from the heat of heavy castings, an evil which can be modified by mixing the sand with coke-dust, or anthracite powder. In tertiary regions, and along the sea-coast, some spot is always found where fine and strong sand may be dug; in these localities the best kind is frequently deposited. The greatest difficulty in obtaining sand of a good quality, is mostly encountered in limestone and volcanic regions, also where porphyry, mica slate, and micaceous rocks predominate. Sand which contains too much iron or lime, or still worse, mica, will not adhere, and is liable to absorb and retain too much moisture, and cause rough and unsound castings. Good moulding sand has in its green state a yellowish earthy colour, balls easily on being squeezed in the hand, and, if sufficiently fine, assumes the finest impressions of the skin without adhering to it. White or gray sand is generally either too strong or too weak. Sand for undried moulds—green sand moulds—is generally more open or porous; it should not contain as much clay as that used for dried moulds, or it cannot assume or retain the finest impressions of the pattern. Sand for dry moulding is frequently of the finest kind, and very strong; for heavy castings a coarse but adhesive sand is mostly selected.

Core-sand.—The material most difficult to obtain is

good core-sand. Core-sand should be coarse, very porous, but still very adhesive. Rock-sand—the debris of abraded rock—free-sand from river banks or from the sea-shore, pounded blast-furnace cinder, and other kinds of coarse sand, are frequently mixed with fine strong sand, or with clay; the use of the latter, however, is to be very limited. The best core-sand is frequently found on hillsides, or the very top of hills, in places where feldspathic or primitive rock has recently been decomposed, where the rock contains sufficient clay to make it adhere, and where the coarse angular grains have not supported vegetation, and it is consequently free of all vegetable or animal matter. Where sand of abraded rock cannot be obtained, free-sand, or, which is preferable, pounded blast-furnace cinder may be used, tempered with clay, barm, pease-meal, or horsedung. In the use of the latter vegetable and animal substances, caution is to be exercised to prevent the boiling of the casting, because of the quantity of gas liberated from such matter. For cores, fresh sand must be used in each cast; old sand, burned sand, or sand mixed with coal, cannot be employed for this purpose.

Clay is frequently used for improving the adhesiveness of sand. It is to be selected from the white aluminous kind, argillaceous earth, or fine clay. It is either dissolved in a large quantity of water, and kept in the foundry for occasional use, or is dried, pounded, run through a fine sieve, and then mixed with the sand. The best plan is, to mix sand and loam together, and run this mixture moist through a mill; a common grist-mill, or a dust-mill, will answer for this purpose. One part of clay mixed with nine parts of free-sand, or any other pure sand, is considered sufficiently strong for core-sand; still these proportions depend very much on the nature of the sand, and the adhesiveness of the clay, and also what kind of cores are to be made from it. The sand for large and complicated cores, is to be stronger than that for small cores.

Loam.—Common loam, or clay of which common bricks are made, is generally used for loam-moulding. The loam ought to be as free from iron, lime, magnesia, and other alkaline matter as possible, because they make the loam too hard and close, and cause boiling of the metal. Such mixtures are also not sufficiently refractory to resist the heat of a large mass of melted iron. If good loam cannot be obtained, a mixture of sand and clay, as described above, is preferable to any imperfect loam. Loam, or any cement for loam-moulding, is to be mixed with saw-dust, horse-dung, hair, or cut straw, hay, or similar matter, which makes the loam adhesive and porous.

Coal-dust, black-lead, and *anthracite dust,* are simply means of blackening the mould, by mixing it with the sand or loam. If hot metal is allowed to be in immediate contact with some kinds of fresh sand, the sand will partially melt, or if the sand is coarse, the hot metal will penetrate into the spaces between the grains, and the casting in consequence will be rough. Blackening, or a coating of carbon, will prevent in a great measure the burning of the sand, and consequent roughness of the casting. Black-lead is a very effective material for this purpose; but if used in too large a quantity it is apt to fill the necessary pores of the sand, and, as it is almost incombustible, will prevent the escape of gases from the hot metal, and consequently cause unsound castings. Next to plumbago in refractory quality is anthracite; and its dust, if not too fine, is an excellent means of preventing the burning of the sand. If there is too much anthracite dust in the sand, it will impair its strength; and if the dust is too fine, it will fill the pores of the sand. Dust of bituminous coal weakens the sand considerably, but it makes it very porous and open, thus facilitating the escape of the gas. It causes the castings to be very smooth, but without fine impressions; it entirely destroys the sharp angles. Bituminous stone-coal dust appears to have a remarkable influence upon iron. Cast in a mould composed of sand and bituminous coal, the iron appears to be more gray and coarse-grained than when in any other mould. It is in consequence generally weaker; pig No. 2 improves by it. Coke-dust mixed with sand is better than any of the enumerated materials for making large castings, and for casting stove-plates. It makes the sand open, without impairing its strength too much. Coke-dust is not well qualified for face-dust; it does not make smooth castings. The most generally useful coal-powder is charcoal dust—ground charcoal of hard wood, such as oak, beech, sugar maple, hickory, or dogwood, well burned. Charcoal powder can be mixed with sand to nearly one-tenth of its volume. It is an excellent face-dust for small castings. Very small delicate castings require a very strong fine sand, free of all coal and coal-dust; these cannot be dusted with charcoal or any other dust, for such would impair the finer parts of the mould. Very small moulds are blackened by a rush candle, or the flame of a pine-knot.

Soapstone powder is a very efficacious means

of preventing the burning of the sand. For thin castings, as stove-plates and hollow-ware, it is not excelled in making smooth, sharp castings. Its use, however, is not to be carried to an excess, because it is as weak as coal-dust, and finally spoils the sand of the foundry by making it too weak. Coal will burn out of the sand, but the magnesia of the soapstone will not; both cause porosity, as well as weakness of sand.

Sand, clay, coal of every kind, and blackening are so abundant in the United States, that we hardly need enumerate localities. Good moulding-sand is found everywhere along the eastern slope of the Alleghenies, from the old rocks of Maine, through the metamorphic strata of New Jersey to the Mississippi river, along the sea-coast in the tertiary deposit, or in the coal and gold regions of Pennsylvania, Maryland, Virginia, and the Carolinas. In the coal basins of the Allegheny, Monongahela, and Ohio rivers, there is no lack of good moulding-sand, and the same may be said of the valleys of the Missouri and Mississippi. Clay is also found there in abundance, and of good quality. Anthracite is in Pennsylvania, in Massachusetts, Ohio, and North Carolina, and where it is found, there is hard bituminous coal, or splint coal, which serves the same purpose. Bituminous coal and charcoal are found in every region of the union. Plumbago is found in Pennsylvania, Virginia, North Carolina, and other places. Soapstone exists in Maryland, Pennsylvania, New Jersey, New York, and along the Atlantic coast. There is an abundance of good materials spread all over the United States.

Mills for grinding blackening.—Coal-dust is prepared in mills of a particular construction, to prevent the flying about of the blackcoal. It is commonly ground in iron barrels which turn around their own axis, and in which a number of cast-iron balls roll over the coal and break it, as represented in figure 1. Such an iron cylinder is generally

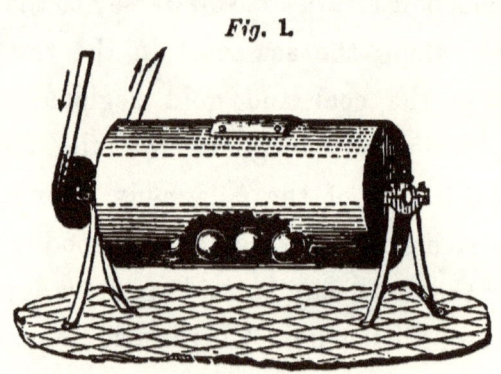

Fig. L.

from 2 to 3 feet in diameter, and from 1 to 5 feet long. It makes from 20 to 30 revolutions per minute, and is moved by a strap and pulley, or cog-wheels. The number of balls, of which each one weighs from 25 to 50 pounds, is indifferent; the more there are at work the better. In the larger cities, as in Boston, New York, and Philadelphia, the manufacturing of blackening and dust is carried on by men who make an exclusive business of it. Remote and country foundries prepare their own dust.

TOOLS.

The instruments and tools used by the moulder are various and expensive. For moulding in green as well as in dry sand, boxes or flasks are used; these may be made of iron or of wood. Iron boxes are in the course of time the cheapest. For moulding in loam, iron plates, core spindles, wrought-iron bars, hoops and wire, are used.

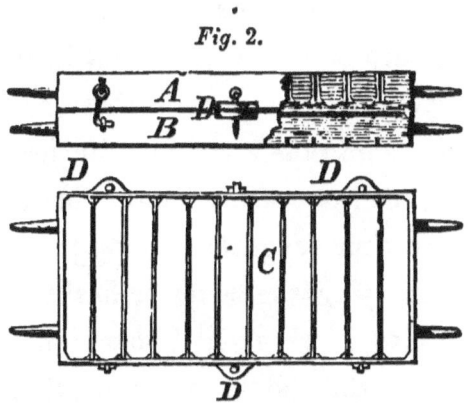

Fig. 2.

Boxes or *flasks* are the enclosures of the sand;

which is filled around the pattern. A flask consists of two parts, as is represented in figure 2, where A is the upper box, and B the lower box. C shows the flask from above. The traverses, which are generally wider in the upper box than in the lower, are best made of wood, even if the box is made of cast iron. These traverses are moveable, and may be put into such positions as to suit the varied forms of the patterns. The three iron pins, D D D, are to be well pointed and tapered, and long enough to afford a safe descent of the one box upon the other. In case there are high projections on the pattern, these pins ought to be nearly as long as the flask itself is high. On each side of the flask are two hooks, fitting to eyes, which serve to connect the two parts of the flask as firmly as possible, to prevent a separation or the lifting of the upper box. These hooks are to be strong without being unnecessarily heavy. The eyes in which these hooks fit, are firmly fastened into the wood and clinched inside, or are cast into the iron when the box is being cast. On each box are four snugs or handles; these are for lifting and carrying the boxes or flasks. On large boxes, and also on very small boxes, there are but two handles, in the middle of the small side, strong enough to bear the weight of the box when filled with sand. In this case the snugs, or swivels, are in the axis of the box; and if a box is suspended by a crane, it may be turned around its swivels, and be at rest in every position. Figure 3 shows a box

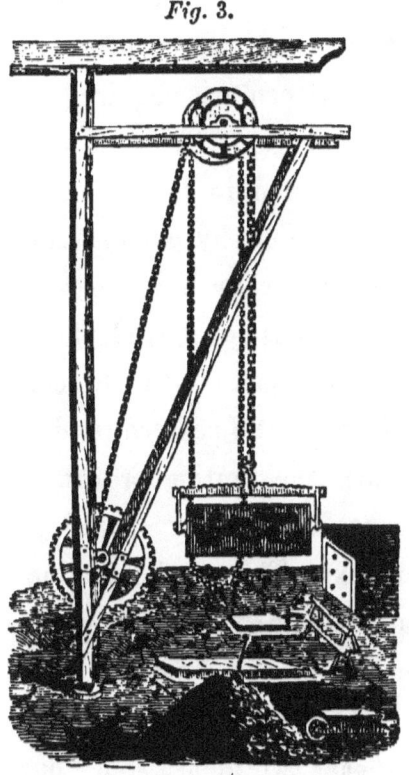

Fig. 3.

suspended from a crane; which in most instances is the proper way of lifting it. We see here that a box must be very strong to resist the influence of the heavy weight of sand and iron. If the box gives way, the sand will crack and drop out, spoiling the mould. Large boxes should always be made of iron. The form of the box is generally suited to the pattern; if the pattern is round, the box is made round. This close fitting of the box to the pattern is in many instances expensive; it causes new boxes to be made where often but one or two castings of a pattern are required. The only inconvenience resulting from square boxes, is the amount of dead sand in the corners of the flask, which may be avoided by putting corners of wood or iron in the upper or both boxes. As in most cases the lower box is not moved, the weight of sand in that part of the flask is of little consequence; but where the nature of the pattern renders it necessary to lift and turn the bottom or drag-box, the corners of a square box may be spared just as well as in the upper box. The chief objection to a square box for round castings, is its weight; but where a strong crane is in the foundry, a little more or less weight to be lifted is of small consequence. In all cases, at least two inches space ought to be between the box and the pattern, and in case of heavy castings, more. This space is also to be larger in wooden than in iron boxes. When the space between the box and the pattern is too small, the mould is liable to leak, the hot metal will flow out if the parting between the box and the pattern is too narrow.

Flasks are to be as rough inside as they possibly can be made, for it is by adhesion chiefly that the sand remains in the box. In large flasks, the adhesion of the sand is increased by driving into the traverses and sides of the box, when the box is made of wood, nails of such a length that the points project on the inside. In cast-iron boxes, nails are either cast in the box, or its inner surface is covered with projections, made by driving the piercer an inch or so into the sand before casting the box; the latter mode is preferable. Nails are inconvenient in many cases, and in all cases troublesome; they frequently cause imperfect castings, as the sand never can be rammed as close where nails project, as where there are none. If the sand is not of a uniform closeness, the cast will be imperfect; for where the sand is too loose to resist the pressure of the fluid metal, the casting will bulge. A better method than the foregoing of making the sand adhere, is to put as many traverses in a box as can conveniently be done, and place them as close together as possible. The interior of the box is made wet, traverses and all, with a solution of strong loam or clay. This loam or clay is put on by means of a whitewash or any other brush.

Moulding-boxes ought to be made of cast-iron; it makes strong and durable flasks. Wooden boxes cost less than those made of iron, but are more expensive in the course of time; they are liable to burning and leaking, and never make correct castings; their pins never fit well, and the wood is apt to warp. Hollow-ware, pipes, and ornaments are to be cast in iron flasks exclusively, or such castings are liable to incorrectness. Iron boxes are more heavy than wooden ones, which is objectionable, but, considering the greater security of the iron flask, the work may be done to more advantage than in wooden flasks.

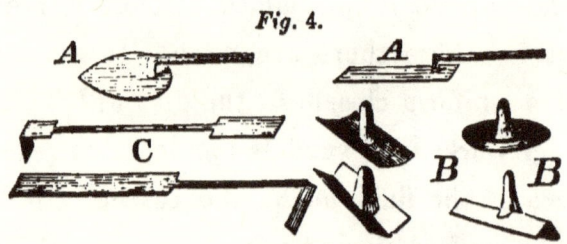

Fig. 4.

Small Tools.—The trowels, Fig. 4, A, A, are from the size of a small mason's trowel, down to one inch long and half an inch wide. The trowel is used for smoothing down the surface of the sand, and clearing away superfluous sand, polishing the blackening or coal-dust, and repairing injuries in the mould. The whole of the trowel is generally made of metal, handle and all. B, B, are round forms of tools for polishing hollow moulds of a cylindrical or spherical form. C is a cleaner, often twelve and more inches long; it is used for cleaning and smoothing sunken surfaces, where the trowel cannot be used. These tools are generally made of steel, but are thus liable to corrosion, which injures their polish. The best metal for tools is hard bronze, as this is not injured by oxidation. A high polish and straight surfaces are the chief requisites of these tools. Their shape or form may be varied, according to individual taste. The general forms as represented, are the most in use.

Fig. 5.

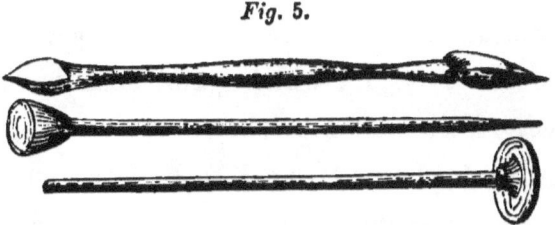

Fig. 5 represents both a wooden rammer and an iron one. The wooden rammer, edge shaped on both ends, is made on the turning-lathe, in one piece; it serves for pressing the sand close into the corners of the pattern, and also into the flask. The other figure represents an iron rammer, which, however, is merely castiron at one end, where there is a round button of from 2 to 4 inches in diameter on the face. The wooden shank or handle is generally tapered or pointed at the opposite end of the knob, for piercing the sand, or to reach more closely into corners. Each of these rammers may be from 2 to 4 feet long, according to the kind of work to be done with it.

Besides the tools here enumerated, the moulder has short-handled light shovels, for filling boxes and for working the sand; sieves of various sizes or meshes, and a riddle for filling the flask; small bellows, for blowing dry loose sand from the mouldings, and parting-sand from the pattern; and also, coal-dust or blackening. The moulder needs an iron pot for holding parting-sand, and also a water-pot; two or more linen bags for coal-dust, black-lead, and pease-meal; a piece of rope for tufts, for which paint-brushes also can be used. Piercers or prickers, are iron or brass needles, made of wire, from $\frac{1}{8}$ to $\frac{1}{4}$ of an inch thick; they are from 6 inches to two and more feet long, tapered the whole length, and drawn to a point.

Parting-sand, is that sand which is strewn over the moulding sand where the boxes separate; it is either free-sand, river-sand, sea-sand, or pounded cinder; or it may be the burnt sand scraped off the castings in cleaning them. *Pease-meal* may be substituted by any other meal; the first, however, is the best. Many tapered pins of various lengths, round, square, oval, and oblong, are needed in a foundry for making gits or gates; some strong, well-tapered and pointed screws for lifting out the patterns; iron hammers and wooden mallets, small crowbars, pinchers and tongs.

Moulding in green-sand.—There are three distinctions in moulding; green-sand, dry-sand, and loam moulding. Green-sand moulding is generally applied to light iron castings; as small, unimportant parts of machinery, stove-plates and stoves, hollow-ware, grate-bars and fire-grates, shot and cart-wheel bushes, water-pipes, gas-pipes, and many other articles. This method is seldom used for any other metal than iron. In making a mould for a small piece of machinery, say a wheel, in green-sand, the pattern is put upon a flat board, which is laid perfectly level upon the floor of the foundry, or, for small articles, upon a pair of trusses, or a box which contains sand. Upon this board the pattern is laid with its smooth side on the board. If the pattern is divided in two halves, but one half of it is laid down, the jointed side upon the board. Figure 6 shows the arrangement seen from above. The board is to be straight and well planed, and

Fig. 6.

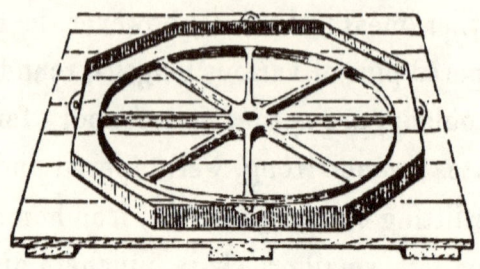

made of two-inch pine plank, or, if the article is small, but one-inch. After the wheel is laid down and well adjusted, or made solid by sprinkling some sand on those places where it does not touch the board, the lower box of the flask is put down inverted upon the board. Before the drag-box is put down, a layer of sand of one inch thick is frequently spread over the pattern and the board. In this sand the box is imbedded, and rests more firmly in it than upon the bare board; the box and pattern are not so liable to shake, or the board to vibrate. The first layer of sand upon the pattern is to be worked through a fine sieve: this sieve is to be finer, the smaller and thinner the pattern, or the more smooth the surface of the casting is to be. This facing-sand, or the first layer, is, in instances where a very smooth sharp impression is required, to be fresh sand from the pit, which never before has been in a mould. Of such fresh sand, a layer of $\frac{1}{8}$ to $\frac{1}{4}$ of an inch in thickness is to be sifted over the pattern. One inch, or, according to the pattern, a greater depth of fine sand, is to form the facing of the mould. All coarse grains of sand are to be prevented from coming in contact with the pattern. If the pattern is complicated, or contains many nooks and corners, the facing is pressed to the pattern by hand, to secure a uniform covering and a uniform tightness of the sand. After the facing is properly secured, common moulding-sand is thrown into the box through a coarse riddle, flush with the box. This sand is rammed down, cautiously and uniformly, with the wooden and edged stamper. When the first box-full of sand is secured and well worked into the cavities of the pattern, the box may be filled again by throwing in sand from the pile, which is repeated until the box is properly filled and of uniform tightness. The coarse, or last sand, is rammed with the round iron stamper, the superfluous sand is stricken off by running an edge rule over the box, so as to make the sand perfectly flush with the box. If this first, or the drag-box, has traverses, as shown in the drawing, there are often difficulties in getting the sand properly distributed over the pattern, and it is not easy to obtain a uniform compactness of the sand. Traverses in the drag-box are admitted only in cases of very smooth single patterns. Most of the moulds are made without traverses in the lower box; it is considered more safe in working the sand, and the work is done easier and faster. When there are no traverses in the lower box, the sand, after being levelled, is sprinkled over with some loose sand and covered with a board, which covers the box all over; it is gently rubbed on, and the whole, box and board, turned over, so that the former bottom is now the top of the box. If the patterns are large, and the box is heavy, it is necessary to fasten both bottoms to the box by means of glands, so that no slipping of the boards may happen while the box is turned over. If traverses are in the box, and no bottom is used, a smooth place on the floor of the foundry is to be prepared beforehand, upon which the box is laid. In case there are no traverses, it is set upon a plank bottom. When the box is deposited in its proper position, that is, in that place where the casting is to be performed, the first bottom upon which the pattern was laid is removed, in which there is no difficulty, if the bottom is not fastened to the pattern. This bottom is frequently fastened to the pattern, which is done in cases where the patterns are limber; as is the case with light and ornamented railing, ornamented stove or fire-grate plates. In this case a few gentle taps are to be given on the back of the board, either with a wooden mallet where the bottom is of value, or with an iron hammer; these taps will loosen the sand at the pattern, and there is less danger of breaking or injuring the facing of the mould. In this case the join-pins of the boxes are fastened to the drag-box, and are to go through the bottom to secure the exact position of the pattern in the sand, when repairs are to be made to the mould, in which cases the pattern is put in again after having been removed. In ordinary cases these pins are fastened to the upper box. In many instances no bottom for the pattern is used, but the upper box of the flask is filled with sand, rammed in and levelled; upon this the pattern is bedded, then the drag-box put on, and the work done as described above. It is a bad practice to work without a pattern-bottom; it is a slow way of working, the patterns are liable to be injured or bent, and the castings are never very fine or correct. After the bottom is removed, the upper surface of the sand-parting is smoothed down, and the superfluous sand cut away by means of a trowel. Pattern, sand, and box are to form one flush surface; this surface forms the parting. The parting-surface is thinly covered with parting-sand, gently sprinkled on by hand; as small a quantity as possible is to be used, just enough to prevent the adhesion of the moulding-sand. As it is impossible to avoid throwing some of the parting-sand on the pattern, which, if left there, would cause a rough surface to the casting, this sand is gently blown off the pattern with a small hand-bellows. After the one half of the mould is so far prepared, the other parts of the pattern are put on, in cases where the pattern is divided; the upper box is then laid in its proper place, the hooks fastened, the facing-sand is put on; after which the common sand is stamped in; in short, the same operation is performed as previously described for the lower box. When the pattern is simple and smooth, there is not much difficulty in adjusting the traverses, which may be straight, and reach with their lower edge down to within half an inch of the pattern. If the pattern is not smooth, and parts of it project into the upper box, the traverses are to be cut out in those places where they touch the relief parts of the pattern. For these reasons wooden traverses are preferable to iron ones, because they can be easily fitted to any pattern. Many boxes have no traverses at all; this is the case with boxes of less than eighteen inches or two feet square.

Gates.—Immediately after the face-sand is put in

the upper box, and before the second layer is thrown in, preparations are made for the *gits*, gates or passages for the metal. This is done by setting in wooden pins, very much tapered, and of a sufficient length to reach above the edge of the upper box. These pins are generally made of wood, and are of a great variety of forms, lengths, and thicknesses. The setting of these for gits is a nice point, and requires some discrimination on the part of the moulder; particularly where iron is to be cast, and where the patterns are very thin. On the distribution of the gits depends in a great measure the success of casting. If the pattern is of a heavy thick form, say more than half an inch thick in its thinnest parts, and its surface is not too large, one gate will be sufficient. In proportion as the surface increases or the pattern is thinner, the number of passages is to be increased. In most instances it is preferable to have the gits outside the pattern; but this always requires a somewhat larger flask, for which reason this rule is not adhered to. Thin plates require flat gits of a very oblong form; mere edges, in case the gits are to be set upon the plate or the casting itself. On round patterns, wheels, pulleys, or any others of that description, the gits must always be set outside. In all cases there is to be an air or gas gate, which is always set upon the pattern directly, whether the passages are inside or outside of the latter. For very light, thin, or open ornamental castings, it is often difficult to find the proper places for the gits, and it requires some experience to decide, at first sight, where to put the gates on a new pattern. Frequently more than one of the first castings of a new pattern are lost on this account. In all instances it is a rule to put the gits in such places that the metal may find the shortest way to fill the mould; where the metal, in passing through the narrowest parts, will find wider and heavier channels to be filled, so that the partially cooled metal may unite again in the heavier parts of the mould. If one passage is not sufficient, there are to be two or more; in fact, as many as are necessary to secure success. The fluid metal is to be poured into all the gits at once, whatever number there may be, so as to fill the mould in the shortest time, and promote a union of the metal from the various passages.

When boxes, pattern, and gits are in their proper places, the flask has the appearance of Figure 7. When the upper box is well filled with sand and levelled, the hooks are unfastened, and the top box gently lifted by one, two, or more men, or, which is

MOULDING.

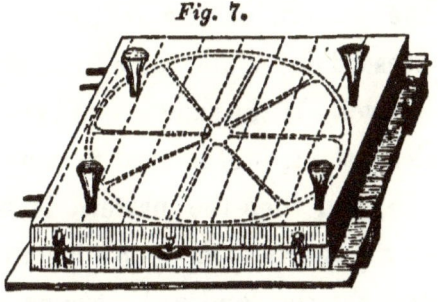

Fig. 7.

safest, by means of a crane. The box is then set on one edge, or turned edgeways in the crane; the pins for the gits are then withdrawn, and the tapering holes are cut larger, bell-mouth shaped, at the top of the flask. The gits are to be very tapering and smooth, to allow an easy passage for the hot metal, and prevent the washing down of loose sand. When the upper box is well mended and secured, and ready to be put on again, the pattern in the lower box is removed. Before this can be done, the edges of the sand all around the pattern are wetted, which is done with a swab, or with a paint-brush soaked in water, and pressed gently between the fingers while running it over the mould. In that way a greater or smaller quantity of water may be thrown on the edges, as the workman may find it necessary. The sand is now examined with the finger all around the pattern, in order to ascertain if it is of a uniform closeness. If too loose, so as not to resist the withdrawal of the pattern or the influence of the hot metal, it is pressed down, and some fresh sand worked in with the trowel. If the sand around the pattern is uniformly close, the trowel is used for smoothing the whole surface, and then the pattern is withdrawn. To withdraw a pattern is in many instances a delicate operation, for the sand will more or less adhere to it and damage the mould, in case the pattern is lifted without being properly liberated from the sand. To free the pattern from the adherent sand, the lifting-screws are put in, after which it is loosened by striking it gently downward with a wooden mallet. In lifting it, it is to be tapped sideways against one of the corners of the pattern, or against the lifting-screws, or against studs made for the purpose.

The lifting-screws are sharp-pointed and tapered, and of a coarse thread when the pattern is of wood. In metal patterns the thread is cut into the pattern, fitting the screw. Richly ornamented or carved patterns, or those of complicated machinery, are seldom lifted without breaking more or less of the mould, and damaging it. The moulder repairs such damages by putting some water on with the swab, and adding as much sand as appears to him sufficient for filling the break. The more prominent parts receive a touch of the swab. The pattern, when removed, is well cleaned by means of a dry brush, and laid in the sand again, in its former bed. With simple patterns this latter operation is not necessary: a skilful moulder can repair a damaged mould without resorting to this expedient. In ornamental moulds there is, however, no chance of successfully repairing a break. The pattern is once more pressed down to its former site, and then withdrawn, the mould generally being then found to be perfect.

Blackening the mould.—By shaking a small bag filled with blackening or ground charcoal, over the mould, it is covered with a thin film of coal-dust. This dust is to be distributed as evenly as possible. If fresh sand has been used for facing, the dust will adhere to the sand, and the pattern, after being well brushed over, may be laid in again to smooth the dust down. The sand around the pattern is smoothed with the trowel. If the mould is faced with old sand, the dust is not likely to adhere, and may be blown off, which is to be avoided. In this case a coating of fine meal is given to the mould; any meal will answer for this purpose, either rice, corn, or pease-meal. If meal has been used before the dust is put on, it is not advisable to put the pattern again in the mould, until a heavy coating of dust has been given over the meal. Care must be taken in using coal-dust or meal, as both cause dull castings if used to excess. The best and smoothest castings are made where the facing consists of a thin coating of fresh sand, and with as little blackening as possible brought upon it. Skilful moulders will however succeed in putting in the pattern again, whether they have been using meal or not. When the sand is well smoothed down, and the pattern laid in again, the channels or passages are scooped out of the parting surface. The pins which formed the gits, have given an impression in the sand of the lower box. Between these impressions and the pattern, channels are dug a quarter of an inch or more deep: where these channels join the pattern, they are seldom more than of the above-mentioned thickness, but may be thicker and narrower towards the gate; the channels must be thinner at the pattern than anywhere else, to make them break close to the pattern, when broken off. If one of such channels is not deemed sufficient, two or more may be cut from the same gate; the channel also may be widened towards the pattern, to afford a sufficient inlet for the metal, and may be swabbed, to give greater security against being washed away by the hot metal. After this is done, the pattern is taken out once more, the upper box

MOULDING. 43

put on gently, the hooks fastened, and the mould is ready for casting.

When parts of the pattern project into the upper box, or the pattern is divided, the same process is to be followed with the upper, as has been done with the lower box. In this case the upper part of the box is to be covered with a board after the gate-pins are withdrawn, and the box laid upon its back, so as to have that part of the pattern uppermost, which is to be withdrawn. The process of lifting the pattern is here exactly the same as in the lower box, except that more caution is required in patching up damages than in the lower box, to prevent the dropping of sand when putting this box on the other.

When a pattern is fastened to the pattern board, it is lifted out before the upper box can be filled with sand. In this case the upper box is filled over a smooth board, well polished with the trowel, and put on without further preparation. It is preferable in this instance to bear the upper box down by weights of pig-iron, instead of hooks. This mode of moulding is easy and works fast, but is only applicable to very tapered and low patterns.

Composition of Moulding-sand.—Although moulding in green-sand at first sight appears to be so simple, yet great difficulties, and often failures, may be encountered by not observing certain practical rules. The composition of the moulding-sand is of the first importance. If the sand is too strong, that is, if it contains too much clay, it is only fit for small or very thin castings. In this instance, care is to be taken not to make it too wet, for it absorbs a great deal of water without showing dampness, but it is soon found to be too damp for casting. Such fat, strong sand may be improved by burning it, or by continual use. It may also be improved by a mixture of charcoal-dust, coke-dust, or anthracite-dust. If too much coal-dust is used to make the sand work well, the castings are apt to be rough. Such strong sand is to be avoided for heavy castings. The heavier the cast, the poorer and coarser the sand is to be. Fine moulding-sand is liable to the same objections as strong sand; it works well in small moulds, if mixed with charcoal-dust, but it will not do for heavy castings. A large mass of hot metal generates a great quantity of steam in the moist sand, also compounds of carbon, which gases require vent: open coarse sand is necessary to give that vent. Core-sand is always coarser then moulding-sand, and seldom fit to be mixed with it. Where many cores are used, whether large or small, it is advisable to carry the castings to some spot in or out of the foundry, where the cores may be withdrawn and broken without their sand mingling with the moulding-sand of the foundry. A lot of good, well prepared old sand, is of great value in a foundry; its proper aggregation ought to be kept up by daily additions of fresh sand, or is liable to become too weak in the course of time. After each casting the sand is to be wetted with as much water as is required to give it the dampness necessary for its adhesion. The amount of water differs in almost every instance, and can be determined only by experience. All the sand of a foundry ought to be riddled at least once a week, to free it from chips of wood, pieces of iron, lumps of burnt sand, and similar matters, which produce inconveniences in founding. If green sand is rammed too tightly, especially for large castings, it is frequently broken, and bad porous castings are the consequence. This happens because the confined steam or gases cannot escape through the sand, and in rushing over the face tear it down. The running in of the piercer, to make artificial air-holes, is in such cases of great service, but is almost ineffectual in large or thick castings. It needs open, porous sand, to make the best kind of vent. Vent-holes pierced or left purposely, will never replace the advantages of open sand. If the sand is not rammed tight enough, the liquid metal is apt to break down all the projections in the sand, and by its fluid pressure cause unevenness and swelling of the mould, and in consequence imperfect castings. Each kind of sand, and each form of pattern, requires a different treatment to insure success. Too loose open sand, and too much coal or blackening, will make rough, imperfect, dull castings. Fine or strong sand is liable to cause boiling, explosions, or porous castings. Many of the difficulties may be removed by a skilful moulder; still it cannot be expected of him to make smooth sharp castings in coarse sand, or in sand which contains too much coal. The skill of a green-sand moulder is more frequently put to the test, than that of any other artisan. Every different form of pattern, different sand, different coal, different metal, and different locality, makes it necessary to modify his mode of working

Division of labour.—The most successful way of overcoming the practical difficulties of green-sand moulding, is to divide the business into branches, so that each different kind of casting may be carried on in its own appropriate locality, and with its own proper workmen and materials. The sand suitable for heavy machine castings, is not fit for moulding small cog-wheels, less so for hollow-ware, and still less proper for ornamental carved castings. The moulder who has been trained to small articles, is hardly able to do heavy machine work; and those moulders who have been used to moulding heavy articles, cannot at once compete with moulders of light castings. To work successfully in green-sand, it is almost absolutely necessary to divide the articles of manufacture. There ought to be a separate shop, and separate hands, and particular sand for heavy machine-frames; a division for small machine-castings; a separate foundry for hollow-ware and stoves; and another for casting ornaments and railings, for brass and for bronze. Each branch of these articles of founding requires peculiar conditions under which it can be most perfectly done, and carried on with the largest profit. The author has observed an instance where a moulder had been making, for eight consecutive years, a certain kind of flat-bottomed pot, with great success. No other moulder could earn half as much on the same article, nor make it equal in quality. This moulder could not make anything else but that pot; he failed in everything else he tried. Moulding generally is a very particular art, but green-sand moulding more so than any other kind of moulding, if we wish to economize in the prosecution of the business.

Moulding in open sand is frequently resorted to, to avoid the making of flasks. It is in no way cheaper than moulding in boxes, and the castings are always rough and uncouth; but there are instances where it cannot be avoided. To mould in open sand, a particular bed is prepared in the foundry. The ground below it is dug out to the depth of two feet below the level of the foundry floor. This hollow is to be as large in extent as the largest mould to be made; a little larger does no harm. It is filled with coarse charcoal, coke, or anthracite-dust, or even with small, say half-inch pebbles, in the bottom. Upon this bed of open matter, two inches thick of the coarsest mould, or river sand, is riddled, and upon this common moulding-sand is sifted. When the bed is so far prepared, two straight edge-rules are put edgeways, one on each long side of the bed. These rules are adjusted by a level, so as exactly to range with each other, as well as with a horizontal line. If now an edge-rule is drawn slanting over these edges, it of course will cut the sand between the rules down where it is too high, and will fill any cavities there may be. As this surface of the sand will still be rough, even after this levelling is accomplished, some fine sand is now sifted over the whole surface, and a long straight wooden roller, of about six or eight inches in diameter, and long enough to reach over both edge-rules in the ground, is rolled gently backwards and forwards over the bed, care being taken that the edges of the rules are clean, and that the roller never misses them. This operation will smooth the surface of the bed; and in case the sand is not considered sufficiently solid, some more fine sand is sifted on, and the roller used to press it down. This process may be repeated as often as it is found necessary, until the sand is sufficiently compact to resist the pressure of the fluid metal. After finishing the bed, the rules are removed. Upon this level bed the pattern is laid; if it has any projections, these are turned downwards and pressed into the sand; the largest part of the pattern however is left above the sand, particularly if the pattern forms a plate. Around the pattern, which is to have a straight surface, some sand is piled by hand to form a dam all around the pattern, and flush with it. After the pattern is withdrawn the sand-dam forms the enclosure, and must be strong enough to resist the pressure of the fluid metal. On a convenient side of the mould the channel is elevated; that is, a place on the top of the dam is made broad enough to receive the fluid metal, and distribute it gently over the mould. If there are any cores in the mould, these are to be held down by pieces of iron, to prevent their being lifted by the fluid metal. After casting, the hot congealed metal should be covered by a thin coating of sand, to prevent its radiating too much heat into the work-room. This kind of moulding is hardly ever used but for the roughest kind of iron castings; it is seldom applied to other metals. It is mostly in use for foundry utensils, as plates and platforms for the loam-moulder, furnace-plates, grate-bars, and the like articles. Plates of any size and form may be made without pattern: the edges are then formed by rulers, and the corners by wooden squares of the desired angle. The thickness of such plates is determined by the amount of metal poured into the mould. Rough flooring plates, rough railing, and other indifferent castings, are sometimes made in open sand.

Moulding in one box.—In castings which are to be made from smooth patterns, and where no great accuracy is required, the pattern may be sunk into the foundry floor and covered by a box. Every foundry floor is considered to consist of sand, at least a couple of feet deep. A ditch is dug, or a place as large as the pattern, and every coarse piece of burnt sand, nails, iron, &c., removed, by riddling the sand. If the place is too dry, some water is thrown over it, and if too damp, dry sand is thrown over until it is so far elevated that the moisture will not injure the casting. The place is to be level. The pattern to be moulded is laid upon the sand and pressed into it, and the sand worked against the pattern by hand. The filling-up around the pattern is to be flush with the pattern, and to extend far enough to resist the pressure of the fluid metal. Upon this mould, which forms the lower box, the upper box is laid, and kept in its place by four or more wood-poles, driven around the box into the ground. This upper box is managed just as any other upper box, with only this difference, that weights are used to bear down upon it and resist the fluid pressure of the metal. If a pattern is large, and there are no means in the foundry to lift a heavy box, and if the upper side of the pattern is smooth, the mould may be covered with iron frames in the form of open network, cast in open sand, and covered with a coating of coarse loam, well dried. By joining the edges closely where these plates meet, a casting may be made just as good as if an upper box had been used. Castings made in these kinds of moulds are never so good as if made in the regular way in two boxes; moulding in this manner is admissible only where necessity compels, and quality is no desideratum. It is in rather more general use than there is need for. In a foundry where large machine castings are made, it requires much room and considerable dead capital to keep a sufficient stock of flasks, but the interest on capital thus invested is easily paid for by the facilities and security afforded in moulding, and the better quality of the castings. Moulding in the floor of the foundry answers for some kinds of pig-iron better than for others.

Moulding of a Cog-wheel. — Heavy green-sand mouldings are very frequent, and it will not be amiss to describe the moulding of a large piece. We will select the moulding of a large face-cogwheel. Some of the wheel-patterns are divided into arms and circumference, which is on many accounts preferable to other methods, but particularly on account of exactness. A wheel cast to its spokes is never round, as the arcs between the arms stretch in cooling. We will adopt a wheel with arms, and these arms divided on account of their cross section.

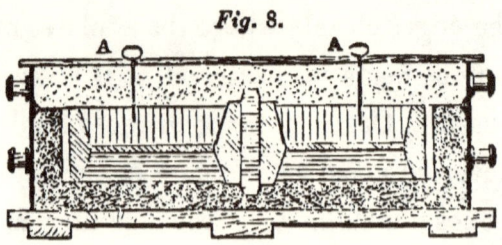

Fig. 8.

Figure 8 is a vertical section of a flask filled with

sand, and ready for lifting the upper box. The different shades of the sand indicate what belongs to the upper and what to the lower box. In a wheel of this kind the face of the wheel is square, as a matter of necessity; no tapering is permitted, as in patterns of other descriptions. The inside of the rim may be tapered, and as the spokes of the wheel cannot be lifted from the lower box, only the spokes are divided so as to lift one half of each with the upper box. The lifting of the upper box is now not difficult, since a part of the pattern is carried with it. The part of the pattern which belongs to the upper box, is fastened to the box by the screws A, A, which pass through the sand, and are fastened to planks on the top of the box. These screws are drawn tight, so as to leave no space for any motion of the pattern. The half pattern in the lower box is withdrawn, by lifting it perfectly vertical and in all its parts at once. This work is done by several men; ten or more hands are often required to perform this part successfully. While the pattern is being raised, the men lift with one hand on iron pins firmly screwed into the pattern, and strike the pattern gently but in rapid succession, so as to loosen the adhering sand. Before the pattern is lifted the damages done by removing the upper box are repaired, which is easily accomplished by using some damp sand and the trowel. In case the sand is not very porous, it is pierced close to the pattern, to make air holes for the escape of the gases. The number of holes required depends entirely on the quality of the sand; close, strong, or fine sand requires more vent-holes than that which is coarse and open. If the pattern in the lower box is smooth and varnished, the swab may be used liberally, but if not, or if the wood is porous or coarse, but little water is used, and the pattern is to be withdrawn as soon as possible. It is altogether a good rule in moulding to work fast, and withdraw the pattern from the sand as soon as possible, particularly a wooden one. It is no advantage to a metal pattern to remain long in the sand; no pattern ought to remain there over night.

It is almost unavoidable to prevent injury to the mould, particularly at the periphery of a cog-wheel; the sand between the teeth will be always more or less broken. To repair these injuries, one or more single teeth are generally supplied by the pattern-maker, of which two at once may be set in and the sand between the two filled up by means of a long sleeker. A preferable mode is to have a segment of the wheel, of at least three teeth; such a segment may be easily withdrawn, and gives more correct divisions. To work with loose teeth requires great experience not to injure the division or pitch of the wheel. Other parts of the mould are generally simple, and if any injury is done it is not difficult to repair such with damp sand, by means of the trowel or sleeker. A long, well made, and polished sleeker is of great service in moulding wheels. The mould is well polished over, after the pattern is withdrawn and every broken part mended; it then receives a slight sprinkling of charcoal-dust, and is again polished.

When the lower box is finished, the upper box, which is still fastened to and suspended in the crane, may be turned over and laid upon its back. If the box is too heavy, or the means insufficient to turn the box, it is left suspended in the crane as it is, face down. Some temporary supports however ought to be erected below the box, to hold it in case the chain of the crane should break, which would endanger the life of the workman engaged in repairing injuries. All the work to be done at the upper box is in this case accomplished from below the box. While one workman is below, first mending and wetting, and then watching the mould, others unscrew the pins from above, and in case there is any danger of sand breaking loose, the unscrewing is stopped, and the doubtful places soaked with water, and firmly pressed. In many instances hooks of small wire, wet in clay-water, are stuck around the edges of the pattern in the sand. The pattern, after every injury has been repaired, is removed, the mould polished, and the upper box is then ready to be put on the lower. In this instance no coal-dust can be used in polishing the mould; the casting, therefore, will be rough at the upper side. In all cases of divided patterns the better plan is to turn the top box upside down, which gives an equal chance to the upper as to the lower box; the proper work can then be performed on it. To turn a box upside down, requires a suspension of it on two points or swivels; the box must of course be strongly made. In lifting, too much attention cannot be paid to the uniform and vertical raising of the box; the least twisting of it will break the sand and cause injury to the mould. Boxes made too weak are very apt to bend, and often cause the falling out of the sand altogether. After the upper box is well repaired, the gits ready, and the channels cut in the lower mould, the flask may be closed. Hooks are useless on large boxes; the only means to keep the upper box down against the pressure of the fluid metal, is by weights or screws. Planks are laid over it to prevent damage to the mould, and the weight, which may consist of broken pig-iron or any other heavy metal, is gently laid upon these planks; in this way the pressure is more uniformly distributed. The gits to a wheel should be between two spokes, near the periphery, and two or three channels cut from each git, either to the spokes, or, preferably, to the spokes and rim. For a large wheel there are to be at least two gits—three would be better. There are also to be some flow-gates, one in the centre and two or more at the circumference. The gits should to be large, say two inches wide, and also have a wide trumpet-shaped mouth. The channels which conduct the fluid metal from the gits to the mould, are to be smaller in section than the git; for in pouring the metal the git is to be kept full, to avoid the passing in of impurities, as coal, dross, or sand, which may float on the metal; such impurities would injure the casting if permitted to pass into the mould.

Failures from some unforeseen difficulty frequently take place in the moulding and casting of large patterns. Fine strong sand is never to be used for heavy mouldings in green-sand; it invariably causes boiling, or at best, causes the castings to be porous and full of holes. If fine sand is mixed with much coal-powder, it is liable to be too weak to resist the pressure of the metal, or even the drawing of the pattern. It requires too much coal to make fine sand porous enough for heavy castings. Coarse open sand is the best for heavy castings where a large quantity of metal is poured in a mould; such sand however makes rough castings, which can be remedied in various ways. The mixing of coal-powder with coarse sand is not to be recommended, for it makes the sand too weak, and causes the generation of too much gas. Open porous sand, free from coal, can be used to advantage, if the pattern is covered with a layer of fine sand, say one quarter of an inch thick, or such thickness as is sufficient to resist the pressure of the iron; a very thin coating is in most cases sufficient. Such a coating of fine sand, well dusted and polished, will make smooth castings. Coal is not of much use in sand for heavy castings, for if the iron retains its heat long, as it does in ponderous masses, it destroys the coal nearest to it, in consequence of which the casting assumes a peculiar roughness. The only coal which resists the influence of hot iron in large masses, is plumbago or anthracite, but these, if they are so fine as to make a smooth surface, are too fine to admit the free escape of the gases, and if such carbonaceous matter is coarse, it causes as rough castings as coarse sand.

In practice, coal mixed with the sand is advantageous, but it is not to be in excess, and coke or charcoal-dust are to be preferred on account of their peculiar porosity. But in heavy castings, coal can never prevent the metal from penetrating between the grains of sand; and if coal is of no service on the facing, it is of none in the body of the mould. Heavy castings are therefore best made in dried sand or loam, as we shall hereafter describe. Machine frames of a large body of metal, or of little importance, may be moulded in green-sand; but frames which are to be strong, wheels, or beams, ought to be cast in dry sand, for the unequal shrinkage of iron in wet sand, caused by the moisture, is very apt to impair the strength of a casting.

Mouldings of more than two boxes, are not so frequent, and are generally avoided in moulding machine frames. Many a complicated pattern may be moulded in two boxes, if properly managed. If no division of a pattern can be devised to meet all the difficulties, the moulding with cores is resorted to, to meet the emergency. We will illustrate this in one instance. Figure 9 represents a flask in which a pulley is moulded. The pattern of the pulley is divided at the dotted line. After the lower box is filled and turned, the sand is cut out around the

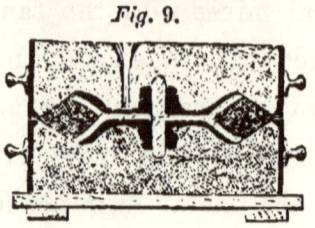

Fig. 9.

circumference as indicated, the surface of the sand smoothed and parting-sand sprinkled on, which is carefully brushed or blown off the pattern. The other or upper part of the pattern is now laid down, and a core of fresh moulding-sand pressed carefully into the groove of the pulley, in the form as indicated. This core is filled flush with the pattern, and slanted towards the edge of the box. It is well polished, covered with parting-sand, and then the upper box put on and moulded. When both boxes are filled, the flask is covered with a board and turned upside down, the drag-box is then lifted off first, and the lower half of the pattern removed. After this the flask is once more closed and turned, putting it this time on its bottom part. The upper box is now lifted, and the other half of the pattern removed. While turning the box, and lifting the pattern, the very brittle round core of green-sand is here always supported, without danger of its breaking. In a similar manner many complicated patterns may be moulded, by simply putting in cores of this kind. Where green cores cannot be applied, dry cores must be used, and the spaces for such provided for in the pattern; but of these hereafter.

Small articles of machinery require in many instances very skilful workmen, and a dexterous handling of the patterns. There is no branch of mechanics where more perfect castings are required and made, than for spinning machines. These castings are to be true, smooth, sound, and malleable, conditions which are not easily effected. To succeed well, it requires particular sand, and a certain amount of coal mixed with it, and workmen who are experienced in that kind of work. Many advantages, however, may be given to the moulder in the arrangement of a pattern. If a small face-wheel is to be moulded, and the teeth are to be parallel, it is difficult to mould such a pattern. If however a ring of lead is cast around the wheel, so that each space between the teeth of the wheel is occupied by a lead tooth, and the wheel may be drawn through the lead without difficulty, the moulding of such a small wheel is rendered comparatively easy, by laying the lead ring upon the sand around the wheel, when the weight of the lead will hold the sand down, which otherwise is apt to follow the wheel, particularly that portion between the teeth. In moulding small machinery of iron, it is not so much the smoothness of the castings which is to be considered, as the soundness of the metal; for this reason, the sand of such a foundry will bear and requires more coal-dust in admixture, than other foundry sand.

Ornamental Moulding.—The moulding of ornaments and railing is a subject of some interest, besides being a branch of the fine arts. Railing of simple forms, with one side smooth, may be cast in open sand; but there is the objection against it that open castings, made of the same metal, are never so strong as those cast in flasks. There is no economy in casting railing in open sand. For coarse railing, open porous sand is used, containing a good portion of coal. Here we have to remember that coal causes faint dull castings; the outlines are generally imperfectly developed. Carved work or sharp outlines can never be expected to be good if too much coal is used, either mixed with the sand, or dusted on. In ornamental moulding, it is not generally the strength of the metal which is the most valuable, but it is the perfect representation of the pattern which is desirable. Sharp outlines and smooth castings are the object of the moulder in this case. Some coal mixed with the sand, is necessary, but it ought to be as little as possible. To secure sharp castings, the facing of the mould is made of fresh fine sand; a layer of one-twelfth of an inch thick is sufficient, and this dusted with fine dust made of oak or hickory charcoal. Ornamental work always is and can be sufficiently tapered to leave the sand readily, and if the pattern is made of metal, and well polished, it may be repeatedly laid in the mould, and all imperfections of the mould may be repaired to the most minute correctness. Dusting the facing of the mould is the very last operation; every damage is to be repaired with fresh sand, and every line of the mould is to be correct before the dust is put on. There is no more coal-dust shaken over the mould, than is just sufficient to make a smooth casting. Pease-meal or any other meal is inadmissible in ornamental moulding; it is injurious to the sharp outlines of the casting. Common pannels of railing are generally smooth on one side, and may be cast in wooden flasks; but where both sides of a railing are ornamented, iron boxes are to be chosen. As an illustration of ornamental green-sand moulding, we will choose a square hollow column or railing-post, represented in figure 10. Figure 11 is the post represented in a section cutting through the post and the flask. The pattern

Fig. 10.

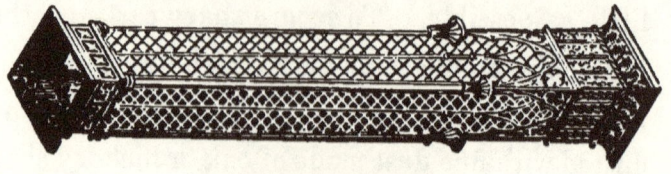

Fig. 11.

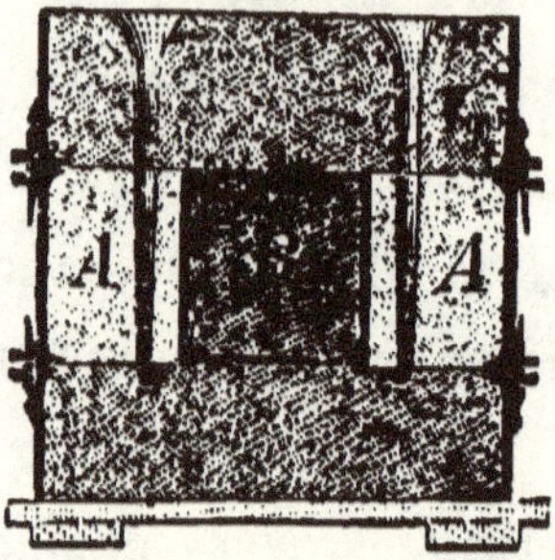

is divisible in four parts; it divides on each corner. In moulding, one of these parts or one side is laid on a board, and the lower box filled over the pattern; the box is then turned, the sand smoothed, and the two other parts A A put on. To keep these parts of the pattern in their places, four or more small square boards are put between them. These boards are of exactly the size to fill the inside, B, of the square. Parting-sand is now thrown on, and the middle box put in its place. The middle box is divisible on both ends, kept together by hooks, so that each part, A, of the box can be removed by itself. The spaces, A A, and B, are now rammed in and filled flush with the pattern and the box. After this the fourth side of the pattern is put in its place, which forms the top; parting-sand and the upper box put on, and this box rammed in. The pins for the gits are to pass through the middle box into the lower; and if the metal is to be not more than half an inch thick in its thinnest parts, it requires four gits, if thinner, six gits. On each end of the column a flow-gate is set upon the upper part of the pattern. When all the boxes are filled, and the gate-pins in, the top is covered with a board, and the flask inverted. The drag-box is now lifted, and the side of the pattern removed. The four parts of the pattern are to be fastened, each to its respective box, by means of screws passing through the sand into the pattern. Each of the four sides of the pattern has its taper towards the box. This lower part of the mould is to be well finished before closing, for there will be no opportunity of getting at it again. The small square boards, B, are now withdrawn, and the spaces left by them in the core, filled up with sand. When the requisite work on this side is performed, the drag-box is put on again and the flask reversed. The git-pins are now withdrawn, the upper box with its part of the pattern removed and put aside, until both parts of the middle box are ready. The pins which hold the middle box to the lower, are not to fit too closely, or are to be moveable, for the parts of the middle box are to be drawn in an angle, because it cannot be done straight. The process of withdrawing the pattern from the middle and upper box is simple, and requires no particular description. For this kind of work a somewhat open sand, or fine sand mixed with ground coke or ground charcoal, is to be used. Too close or too strong sand is liable to cause explosions in this case. Many apparently complicated patterns may, like this pattern, be very easily moulded, and by simple means, if they are properly divided.

Moulding of Hollow-ware.—The distinct objects of this branch are, however numerous, still of great similarity. In no branch of the art of moulding is skill and dexterity brought to such perfection as here; it is the result of the division of labour, practised in this department. The objects belonging to this branch, are pots, kettles, fire-grates, stoves and stove-plates, grate-bars, locks, latches, hinges, and all such articles, which are standard articles of commerce. In this case it is not alone the sharp, well expressed outlines of the pattern which are essential; besides these, well finished articles require smooth surfaces, uniform thickness, and a high degree of lightness. The sand of a hollow-ware foundry is to be fine, but it may be liberally mixed with coal-powder; blackening or anthracite may be used for dusting. The most elegant patterns are now manufactured into stoves, and we may say, that there is no nation where the art of constructing elegant and economical iron stoves and fire-grates, has been carried to so great an extent as in our country. The moulding of these patterns is simple, there are but few complicated forms, and therefore this branch is no particular object of our investigation. In the manufacture of hollow-ware, there is a great advantage in good well-finished patterns. If the patterns are perfect there is generally no difficulty found in making good castings, for most of the articles are thin, and there is little danger of the sand burning and adhering to the metal. Articles of commerce are generally worked to as much advantage as possible. Patterns of small articles, as parts of locks, latches, hinges, knife-blades, knife-covers, and other small articles, are generally put ten or twenty or more together, connected by a permanent channel which conducts the metal from the gits to the patterns, and forms a part of the pattern. Such a batch, or set of patterns, generally fills one flask. The New England States, and Pittsburgh, are remarkably successful in manufacturing small articles. In some cases various articles are put promiscuously into one flask, in which, however, a similarity of size is to be observed. Whatever number of patterns there may be in one flask, it is always calculated to cast a flask of small objects with one ladleful of metal.

Moulding of a Coffee Kettle.—As an object to illustrate hollow moulding, we will choose the form of a common coffee pot, or water kettle, represented as moulded, in figure 12. The form of a water kettle

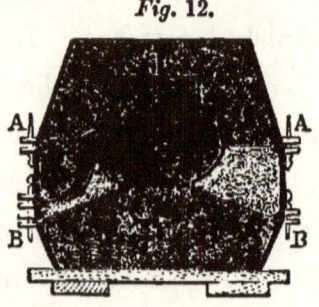

Fig. 12.

is generally known. It is an almost spherical vessel, with a snout or pipe. We have selected one which fits to a cooking stove, with a contracted flat bottom; in other cases that bottom is round, with three studs to stand on. The pattern is here an exact model of the kettle as it is to be, with the exception of the pipe, which is, or may be solid. The flask consists of three boxes, of which the middle box is divided by a vertical division into two halves—cheeks. This division runs through the pipe and divides the mould into two halves, so that when both boxes are removed, the pipe, which is not fastened to the pattern, may be withdrawn. In this case the upper part of the pattern is divided just in the division of the middle box, which leaves an unsightly division, and is likely to expose the pattern to injury. A better plan of working is, to have the middle box in one piece, and divide at the lines A, A, and B, B. At the pipe the upper box reaches down into the middle box, as far as the pipe goes down, and divides the sand just along the bend of the pipe; the middle box parts with the lower at the rim of the kettle, where the core also separates, as indicated by the darker and lighter shades of sand in the drawing. The pattern is only divisible in the line A, A, through the pipe. In moulding this kettle the lower (in the drawing the upper) half is put on a board and the upper box rammed in, this box turned upside down and the other half of the pattern put on. The middle box is then set in its place, and fastened to the upper box. Both boxes may also be put together, and rammed in together, just as conveniently. Sand is then filled in the middle box around the pattern, and after this the sand is rammed inside of the kettle. The parting is made between the lower and middle box, as indicated, and the lower box filled. The flask stands now inverted, and the kettle on its bottom. The lower box—as the flask stands it is the upper box—is now withdrawn, then the middle box lifted and the upper half of the pattern withdrawn. First the middle and then the upper box put on again, and the flask turned, which will now stand as in the drawing. We may now draw the upper box, remove the lower part of the pattern, and put in the core for the pipe, which is made in a separate core-box. The git-pin is now drawn: this is very much tapered one way, and thin, the other way three or four inches wide, formed like a blunt wedge, whose edge is $\frac{1}{4}$ of an inch thick. The box is now put on again, and the mould ready for casting.

Patterns for hollow-ware require to be very accurate, if we expect the moulding to be well done. The originals of these patterns are generally moulded in loam, cast in brass, and turned in a turning lathe, or, if not of a round form, worked by other means until a perfect form is obtained. A pattern having been smoothed and polished, is then cut into such parts as are considered necessary to make it available. Pins, ears for handles, and studs for feet or handles, are generally put on loose. All dished utensils are generally cast with their mouth downwards, except covers. Where the neck of a core is narrow, and there is any danger of the hot metal lifting the core, as may occur in the case of the coffee pot, the core is fastened to the bottom of the flask by a thin iron rod with a cross at the upper end, buried in the core and fastened below the bottom. Hollow-ware moulders need a variety of peculiarly shaped tools, and sleekers. Most of the tools are button-shaped, with short studs for handles, more or less round, or even cylindrical, to suit the various hollow forms of the patterns; others are plain and heart-shaped; others again have double plain surfaces at certain angles with each other, to suit certain corners in the mould. Blackening—plumbago—is chiefly used as dust, and if well polished, it will make smooth and good-looking castings.

In this kind of moulding, iron boxes are generally used; this is necessary to secure good and correct castings; it is also the cheapest way. If iron flasks are well made, the work in them is done fast, well, and safe, while imperfectly made or wooden flasks always cause more or less delay in work. From well made flasks many advantages may be derived: we will mention one. Suppose a moulder is to mould twenty flasks of one and the same pattern, if the boxes are well made and fit one upon the other promiscuously, there is no need of boards after the first drag-box is moulded. Upon the first box which is moulded, say the lower box, its complement the upper box is rammed in. After parting upon the upper box, the next lower box is moulded, leaving of course the pattern always in that box which serves as the bottom of the flask. In this way the top box of the first flask serves as the bottom to the next bottom box, and so on through the whole range of boxes. Each two boxes come together as they have been moulded, and it may happen in the course of the work, that one of the last boxes will not fit to one of the first, which however does not make any perceptible difference in the correctness of the castings. It requires some dexterity and experience to succeed well in this mode of moulding.

There are many articles here not enumerated as belonging to green-sand moulding; such as iron castings, parts of architecture, which are now so extensively used. To this belong window and door sills, door and window frames, columns and railing. All these forms are easily moulded, and require no particular details; we shall, however, mention some of them in the following chapters.

Mixed Sand Moulding.—*Moulding in green sand with dried cores* may be considered a mixed moulding, which requires particular knowledge of the composition and construction of cores. In previous pages we have spoken of core-sand, but we shall here treat upon the formation of cores, and the quality of the core-sand for particular purposes. The management of cores is a matter which requires some ingenuity; malformation often causes perplexing failures, and is in most cases the source of unsound castings.

Cores are especially used in forming vacancies in castings, which cannot be successfully formed by the pattern. The forms of cores vary greatly, as may be expected; but in general, if made of open porous sand, free of vegetable and animal matter, and of coal, and if the sand does not contain too much clay, and the cores are properly dried, there is hardly any difficulty experienced on account of the cores. A caution not to be neglected is, that cores are never to be put into a green-sand mould until the very latest moment before casting. Cores which are not surrounded by metal on all sides, are made of stronger sand than others which are often not dried at all. In cores which are covered with metal on all sides, and have only one or two small vent-holes for the escape of the gases, as is the case with cores for narrow pipes, the sand is moderately mixed with free-sand. It is to have no more clay or adhesive matter than is just necessary to make it adhere for being moulded and dried. Sand of sharp grains, as pounded rock or slag, is more open than the composition of round grains, as river or sea-sand, and for this reason preferable. In many cases, yeast or meal water is used besides clay water to strengthen the core-sand, but these ought to be used cautiously, for not only water, but any other substance which generates gases is injurious to core-sand, causing blower holes in the castings. The safest core-sand is a natural sand which can be used without any artificial admixtures. Moulders ought to examine their neighbourhood until they find sand suited to their purpose, in case they are not already provided with it. Long or thin cores are stiffened by iron wires, or small rods of iron, which are moistened with clay water. Such wires or rods are buried in the core, and recovered when the casting is cleansed from its adhering sand. Cores of considerable length, also those in which the sand is rather strong, are pierced with long wires through the whole length, taking care not to drive the piercer through the surface of the core. Curved or angular cores, which cannot be pierced, and are too long to do without vent-holes, are made open by laying one or more strings along the stiffening wire in the heart of the core, which strings are drawn out after the core is dry. If cores are too long to bear their own weight and the pressure of the metal, they are to be supported by chaplets or by staples. The latter are simply nails with broad flat heads; they are stuck into the sand mould, and project with their heads just so far as the thickness of the metal between the mould and the core is to be. Chaplets are simply bent pieces of sheet iron in the form of a [, or two pieces of sheet iron riveted to a pin, the distance between both being equal to the thickness of the metal. Cores are covered with a coating of blackening, which is put on wet. This is the more necessary, as the cavities made by cores are mostly difficult of access, and an easy scaling off of the sand from the iron is therefore very desirable. Liquid blackening for cores is the same as that used in loam-moulding; and by referring to that chapter a receipt for its composition will be found. The blackening is laid on the wet core, as it leaves the core-box, by means of a heavy paint brush, and both the core and its blackening are dried simultaneously.

Small common cores are made in simple core-boxes, such as are represented in figure 13. A,

Fig. 13.

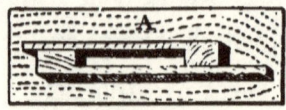
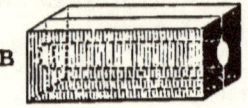

is two strips of boards, with a square projection on each end. Both are at liberty to be moved, and if laid upon a flat board, sand may be filled in the space which is formed by the squares: for each size, that is, section of core, such a box is required, but any length of core of that size may be made in a box of this kind. Round cores are made in boxes similar to that represented in fig. B. Globular cores are made in spherical cavities, and in fact any core in such a cavity as it is destined to form in the casting. Cores are not always made because they are necessary: they are frequently made to save expense in patterns and in moulding, and to render a successful cast more certain.

Moulding of a Column.—As an instance of mixed moulding, we will describe the moulding of a fluted column, which may serve as an illustration for most cases of this kind, particularly for pipes. Figure 14 represents the pattern of a column with orna-

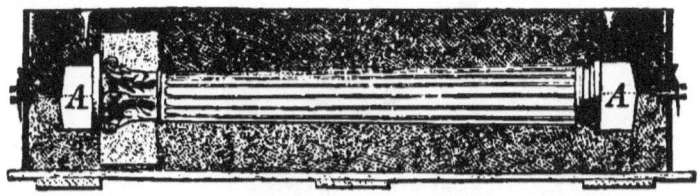

Fig. 14.

mented capital, as it is imbedded in the sand, moulded, and ready for removal. A, A, are the core-prints, which leave a cavity to be filled by the long core which is to form the bore, or hollow in the column. It is in many cases difficult to mould a richly ornamented capital in green-sand, along with the trunk of the column, and still the capital ought to be in a solid connexion with the shaft; this case is here represented. On the pattern of the column, instead of the ornamented capital, a block of six or eight sides, or of more or less than that number, occupying the place of the cap, is inserted, as indicated by the lines in the drawing, and more distinctly represented in figure 15, by the dotted lines which represent a hexagon. The fluted shaft of the pattern is divided through the whole length into two halves, which is best done through the opposite channels, as indicated; for a seam which falls otherwise upon two ribs,

7 *

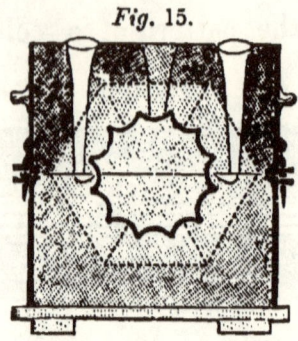

Fig. 15.

makes these ribs always more or less imperfect, which is not so glaring if it falls in the channels. Besides this division of the pattern, each half of the pattern is again divided into three subdivisions, or more, as the case may be. These latter divisions, as shown in the drawing, divide the circumference of the column into six parts, each half in three, held together by blocks and wood-screws. After the screws and the blocks are drawn, the pattern may be taken from the sand in parts, each part by itself. No second part is removed until the first impression is mended in the mould, in case there is anything broken in the sand. The capital is formed in the following manner.: If it consists of six equal ornaments, as leaves or spirals, one of these is carved, and prepared for being used as a pattern; over this pattern a core-box is made, and so calculated that a core made in this box will fill one of the parts of the polygon formed by the pattern. Such a core will fit in the mould, and occupy one part of the space having on one side the impressions of the ornaments of the capital, joining with two sides the next cores, and resting with one side in the sand of the mould. The cores which belong to the upper box may have wires or rods inserted to be fastened with, to the box. After these cores are placed, the centre core is put down, the flask closed, and in fact managed like any other object of our consideration. In placing the cores, care is to be taken that the liquid metal cannot penetrate below a core and lift it; all the crevices which would lead to such a result are to be avoided, or carefully filled up with green-sand; and if there is any doubt as to the safety of the cores, they are to be wired down. At each end of the flask in the parting a small opening is left to communicate with the vent-holes of the core; these openings are in no way connected with the interior of the mould, so as to endanger the cast by admitting hot metal to run out this way. Gits and channels are as usual in the proper places, and if means permit it, the column ought to be cast inclined, into one gate which is at the lowest part, the git raised, by means of small boxes, to such a height as to balance the flow-gate. The latter is to be at the highest point of the pattern and the box. Here, as in any other case, the cast-gate is to be kept full, in pouring in the metal, to prevent the running in of impurities along with the iron. Directly after the column is cast, or better still while the metal is pouring in, fire is to be applied at both ends to kindle the gases escaping from the core, which gases will explode if left to kindle spontaneously.

Water pipes, gas pipes, or pipes for any purpose whatever, are moulded in the same manner as columns. There is no essential difference, but in the form of the pattern. The core of a pipe is to be a fac simile of the bore or hole to be formed. A core-box for water or gas pipes is represented in figure 16: it shows a

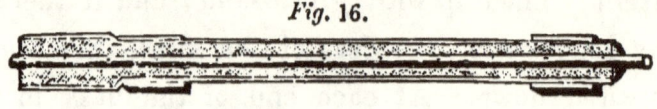

Fig. 16.

longitudinal section of an iron core-box. Frequently such boxes are made of wood; but in well conducted establishments they are made of cast iron. Wood is apt to twist and warp, and in consequence causes imperfect cores. An iron core-box is generally round, about half an inch thick in metal, and has two square projections to rest upon when laid down, around these projections an iron strap is drawn, to hold both halves together when in use. A core-box is to be true in its bore, for which reason it is bored or planed, so as to make it true. Both edges, where the halves of the box join, are, if not quite sharp, to form a blunt edge in case the core is made when lying in a horizontal position. To make good cores in a lying box requires a great deal of experience, and it is for this reason not generally practised. In most cases the box is rammed-in vertical or inclined; the latter way is more convenient than the first, and quite as good. The ramming-in of the sand is done by a long iron ramrod. The centre of the core is, in very thin cores, say $1\frac{1}{2}$ inch diameter, an iron rod, along which a wire is laid; both are rammed in together, and the wire is withdrawn while the core is in the box. This leaves a cylindrical channel all through the core, and serves for the escape of the gases. In thicker cores, of two or three and more inches in diameter, the centre rod is a hollow pipe of cast or wrought iron, full of holes. The latter are necessary, or the gas would not find its way to the interior of the pipe. Heavy cores are made of loam, of which we shall speak in another place. The centre rod is to be a few inches on each end longer than the core. This forms the bearing for the core to rest upon when it is to be dried, and also the journal on which it is to be turned, when the blackening is to be laid on.

Moulding with Plates.—In many cases cast-iron plates with handles are used when one part of the mould is to be removed before the pattern can be drawn. This is the case with the sand between the arms of a bevelled wheel; also with face wheels, or in cases where the pattern, and consequently the sand, is too deep to admit the drawing of the pattern without injury to the mould. Plate moulding is generally performed on bed-plates of steam engines, bed-plates of turning lathes, house props, and in all cases where the sand is surrounded on three sides by hot metal. The sand lifted out in these instances is dried and treated like a core. In the case of a bevelled wheel the moulding by plates is effected as follows: Figure 17 shows a section of a bevelled wheel as it is

Fig. 17.

imbedded in the floor of the foundry, which has been levelled for the purpose. The sand, in immediate contact with the pattern, is sifted. The parting is in the line A, A. In the spaces between the spokes, cast-iron plates, B, B, are inserted, with wrought-iron handles cast into them: these plates are cast in open sand, and from $\frac{1}{2}$ to $\frac{3}{4}$ of an inch thick. They have in this case a triangular form, similar to the space they serve to occupy, and are at least two inches all round, smaller than that space. These plates are laid upon the parting, or, in many instances, impressed into the sand about $\frac{1}{4}$ of an inch deep. They are then covered over with a layer of small iron rods, or wire, or in many cases wooden rods, dipped in clay-water. These rods overhang the plate and reach near to the pattern. The body of sand in the centre of the plate will sustain that end of the rod which is to carry the sand beyond the plate. The space between the arms is then filled upon the plates with moulding sand, flush with the pattern. This forms the parting for the box. After the pattern is covered, and the top box removed, the sand between the arms is removed, by means of the handle C; of which there may be more than one if the core to be lifted is too heavy for one hand, or it is to be lifted by the crane. This part of the mould forms separate pieces: cores in the form of triangles, which may be blackened and dried. The pattern being removed and the other parts of the mould ready for casting, the plates are replaced, either green or dried, just as convenient. The upper box is put on, and the mould may be filled with metal. This kind of moulding is very extensively used; it is a cheap and very convenient way of working.

Dry-Sand Moulding.—This is a very interesting branch of moulding; to it belong most of the brass and bronze moulding, ornamental iron moulding, and a great part of machine moulding. Dry-sand moulding is in many respects preferable to loam-moulding; it gives a casting more true to the pattern than loam, which latter, on account of its shrinkage, frequently gives imperfect forms to the cast. The strength and uniform texture of the castings is quite as well secured in dry-sand moulds as in loam moulds. Dried or baked sand often consists of a mixture of loam which has been used, and fresh sand; in most cases, however, particularly in ornamental moulding, fresh sand is used. Dry-sand obtains a very firm and open texture, and is well qualified to cast machine shafts, pipes, and such articles as require strength and beauty. The manipulation of moulding in dry sand is exactly the same as in green-sand, but is less difficult. In this case no coal powder is mixed with the sand, which leaves the sand stronger. If fresh sand is used, it is of very easy moulding. When the moulds are finished and blackened, they are conveyed to the drying stoves, for at least twelve hours, twenty-four hours, is better to expel by the action of heat the moisture contained in the damp sand. The blackening is done by a paint brush, in the humid way, just as loam moulds or cores are blackened. This is done with some caution, so as not to injure the sharp outlines of the mould. The blackening is applied very thin. A moulder who understands mixing his sand properly, so as to be strong and porous, and assume at the same time fine impressions, will make finer castings in this way than can be done in any other mode of moulding. Dry-sand moulding requires strong iron boxes; wood is impracticable, for even if it did not burn in drying the mould, its warping and twisting would injure the mould. All the traverses, if any are needed, are to be of iron. Long patterns, as shafts, require particularly strong boxes, for these are mostly cast in a vertical, or at least in an inclined position. The pressure upon sand and boxes is then very heavy. Hooks and eyes are in these cases not strong enough to hold the boxes together; it requires *glands* to accomplish this. Glands are double angles, made of flat wrought iron. The rods are to be six inches longer than the height of the flask, or of the two boxes together: these six inches are for bending a square angle at each end of the rod, after which it assumes the form of a [. The distance between the angular ends is to be a little greater than the height of the boxes and bottom. In slanting these glands upon the boxes, the flask may be drawn together as tight as the strength of the glands will permit. The drawing of the glands is performed by small crow-bars gently, so as not to injure the mould. We shall speak of this hereafter. Boxes for dry-sand moulding, if heavy, are to be provided with swivels on each end, for each box is to be turned, the facing of the mould uppermost; blackening and drying require this. In moulding pipes, which need strength, it is necessary to mould them in dry sand, in strong boxes, and to cast them vertically, or at least inclined 30° or 40°. Pipes, or any other objects which are cast horizontal, have always one bad side. The upper side is in most cases porous, unsound, and, in pipes, generally thinner than the bottom side. The liquid metal is apt to lift the core, in spite of staples or chaplets. Another advantage arises from casting vertically, in the better escape of the gas, and the greater security of the core against injury. The core is not so liable to bend and the core-rods may be lighter.

Moulding of a Large Pipe.—There is not the slightest difference between moulding in dry-sand, and moulding in green-sand, except in the composition of the sand, blackening, and drying of the mould; and therefore it hardly seems necessary to illustrate this branch. We will, however, describe the moulding of a large water pipe, as illustrative of this case, and introductory to loam moulding. All water pipes of more than twelve inches diameter, ought to be moulded in dry-sand, and with loam cores. Water pipes are generally made from eight to nine feet long—small pipes frequently but five or six feet long. The pattern is like the exterior of the pipes as it is to be when cast, having at each end a core print five or six inches long. The pattern may be of a solid piece of wood, but is generally composed of strips of plank, to diminish the weight of it; it is divided parallel with its axis, into two halves. After the moulding is performed in the usual way, the mould is blackened and carried to the drying-stove, on an iron tram-road, or by means of a crane. If the foundry possesses no drying stove, or if the boxes are too heavy for transport, some boxes may be put together, a temporary wall of bricks or moulding boxes set around it, covered with sheet iron, and a fire of coke or charcoal or anthracite is kindled below. The boxes are dried in this way on the floor of the foundry. This mode of drying moulds, however, is imperfect, and slow, produces inconvenience in the foundry, and is expensive. The making of a loam core is a matter of no difficulty, if core-bars, loam-board, and loam are in good condition. The core-bar is in this case a hollow, cast-iron, cylindrical pipe, perforated all over its surface, with either round or oblong holes. The core-bar is about three inches less in diameter than the core is to be, with a view to provide room for a hay or straw rope, by which the core is made porous, and so thick as to leave just sufficient space for loam. The core-iron has a journal at each end, made of wrought iron and screwed to the cast pipe, leaving as much opening as possible for the escape of the gases. These bearings, or journals, may be of cast iron, in which case they are made hollow and square inside, to receive a winch by which the core-bar is made to turn upon

Fig. 18.

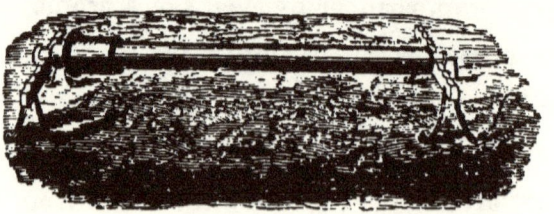

its axis. The bar with its bearings is laid upon two iron trestles, as represented in figure 18, on which it may be turned to receive its hay rope and loam. The trestles are about three or four feet long, and are provided with various sized triangular dentations for different-sized journals. The hay or straw for ropes is kept in a moist place, to have it soft and more fit for being twisted. To make a hay rope, a simple winch, made of quarter inch iron rod, with a wooden handle, is required, such as is represented in figure 19, A. Hay ropes are made by the boys when not

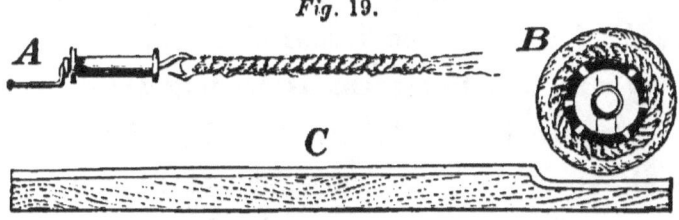

Fig. 19.

otherwise engaged, and kept for use when required. The method in which the rope is applied is simple: the core-bar is laid with its journals in the trestles, as shown in figure 18, the rope fastened at one end, and the bar turned upon its axis; the rope is led so as to make a close and tight covering. If the rope is loose on the spindle, it is liable to be pressed together by the fluid metal, which would, in the most favourable case, injure the casting, but would almost invariably cause a failure of the cast. Wooden core-bars are not to be recommended, as it requires something stronger than wood to resist the pressure of a high column of fluid metal. In figure 19, B represents the cross section of a core, in which the core-iron, journal, hay rope, and loam covering are shown. The hay rope receives a slight covering of thin loam, just sufficient to cover the hay, and remove the roughness of the rope. This coating of loam being dried, the core is taken again in the trusses and the loam-board is applied. The loam-board is, in this case, an almost straight board, of eight or ten inches wide. It is straight every way, and to prevent its bending while in use, it is supported by a rib, screwed to it, or by a strong plank, upon which it rests. The board is so long as to rest upon both trusses, and is fastened to these, just so far from the centre of the core-iron, as to form half the diameter of the finished core. The edge of the board (in the drawing, the upper edge) is shaped as the form of the core, which is in this case almost a straight line, but is cut out, at one end, to form the funnel, or cup-mouth, of the pipe. When the board is in its proper position, and fastened at both extremities by means of weights or screws, it is obvious that when the core-bar is turned upon its axis, it will describe the form of the core along the edge of the board. By turning the core bar with its hay rope and superficial coating of loam, and throwing on additional moist loam, the surplus moist loam will be stricken off by the loam-board, or laid on in those places where the board does not reach the loam. When the core is properly filled up and closely covered with loam, the loam-board is taken away, washed, and put in its place again. The core is now turned somewhat faster than before, and receives a slight washing, merely by dipping the hands into water, and moving them over the surface of the rotary core. When smoothed, which is done with as little water as possible, the core is brought to the stove and dried, then blackened, dried again, and is then ready to be put in the mould. If the cores are long and limber, the staples are not to be forgotten.

The thickness of the covering of loam depends partly on the quality of the loam, but chiefly on the thickness of the metal, and the duration and amount of pressure upon the core by the fluid metal. For common water pipes, if cast inclined, and porous loam is used, one inch is sufficient for the core, but, if cast standing, one inch and a half of loam ought to cover the hay-rope. If the thickness of loam on a core be more than three-quarters of an inch, it is necessary to lay on the loam in two or more layers, always drying the first layer before the next is put on. The loam which forms the core is made as open as possible in its composition; old moulding-sand, old core-sand, or river-sand is mixed with the loam. The working edge of the loam-board is not a square, but is slanted so as to form an angle of nearly 45° to the tangent of the periphery of the core. This is necessary in order to make the board to sleek (to make the loam smooth). A square edge would cause a rough surface to the core. The slanting of the edge is indicated in figure 19, C, at one end of the board.

After careful drying, blackening, and polishing, the core may be put into the mould, if required. The mould is provided with staples so as to support the core, and is then carefully closed. If the box, or the pipe, is large, it is advisable to cover the box by a strong board, and put the glands upon the board, so that there is a board at top and bottom of the flask, to protect the sand from being pushed out. In many instances the moulding boxes are tapered so as to form half a hexagon; in these cases glands are of no use, and heavy iron weights which reach all across the boxes, are used to bear the top-box down. Air-holes at both ends of the box are to be provided for, for the core in this case emits a great deal of combustible gas, which is to be kindled in proper time to prevent explosion.

Pipe moulding is a very common employment in iron foundries, but still there is something peculiar in it, which makes it inconvenient to cast pipes in a foundry where green-sand or dry-sand moulding is done at the same time. It suits best in a loam-moulding establishment. There are great varieties in the form of pipes, but as long as they are straight, a pattern is made and moulded in dry or green sand. The core in this case being also straight, is easily made. It is more difficult to form the core for a bent pipe or knee. We will allude to this in the next chapter.

Casting Pipes without Cores.—There was considerable interest manifested, some time ago, in a process for casting pipes without cores, by means of machinery. An iron mould, well bored and polished, is made to turn upon its axis in a horizontal position; the fluid metal cast in at one end, will naturally flow round in the mould, and if sufficiently fluid, will make a pipe of uniform thickness. How this machine turned out in practice, we do not know, for nothing has been said about it for a long time. Any improvement which will reduce the price of iron water pipes is worthy of notice, and the above machine ought to attract sufficient attention to give it a fair trial. One thing is certain, that every kind of pig-iron is unsuitable for this process.

Moulding of Fine Castings.—Before we conclude this chapter we will give a short description of ornamental moulding; that is, the moulding of small ornaments and trinkets in iron or bronze. There is little difference between moulding for iron, and moulding for bronze; the chief distinction is in the thickness of the cast. Bronze must be cast very thin, if sharp, fine, and distinct outlines are desired. In iron, the same attention need not be paid to the weight of the cast. The principal thing to be attended to in moulding small articles, is the quality of the moulding-sand. This must be as fine as it possibly can be obtained. It ought to have as little clay, or any other foreign admixture, as possible, to prevent its shrinking, and in consequence breaking and cracking. Sand for this purpose is to be an almost pure silicious compound. Coal-powder or any other admixture is inadmissible; it is fatal to the beauty of the cast. Good sand of this kind adheres easily with the least amount of water, takes the finest impressions of the skin, and may be cut into fine slices by a sharp knife. For this kind of work, the greatest evil is too much clay in the sand; other impurities can be removed by sieves, or by washing. Fine tripoli is the best sand for these purposes.

Small articles of bronze or iron, are moulded in the same manner as larger parts of machinery, or hollow-ware. The sand is rammed very close in small iron moulding-boxes, and the boxes dried in the stove, blackened if for iron, but not so if for bronze or brass. For brass or bronze it is advisable to face the mould each time with fresh sand, thrown on through a fine silk sieve. If this coating is but one-twelfth or one-eighth of an inch thick, it improves the casting considerably. Moulds for iron cannot be dusted with charcoal, or black lead, as these would be too coarse. The moulds after they are dry are blackened by a rush-candle, or the black smoke from a pine knot. The box which contains the mould is inverted, so as to turn the face of the mould downwards, rested upon two extreme points. The flame of the candle or wood is held under the mould, which will assume in consequence a velvety coating of fine carbon. There is to be as little blackening as possible; too much will injure the mould and the casting. To mould a simple rosette, or anything which gives but a simple impression in the lower and upper box, is of very easy performance. The case is different with more complicated forms—articles which can not be screwed together, but must be cast in one piece, as statues, columns, and other similar objects This is an interesting art, and it may be of some use to illustrate a few cases of this kind.

Moulding of a Stag.—If the small form of a stag, figure 20, resting upon a platform, is to be moulded,

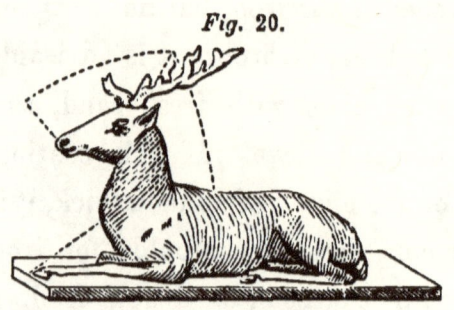

Fig. 20.

it is at once visible that the antlers cannot well be brought into the same mould with the body: they are moulded by themselves, and screwed on. The platform can be cast with the body, but it makes less work in moulding to cast them separate, and screw the platform also to the body. We have now only the body of the animal left to make a mould for. In this case the natural parting is over the back, following the spine, and so dividing the face and breast. The pattern is accordingly cut in two in this line. When one half of the pattern is moulded, the box is turned up, and so much of the sand as cannot be lifted, is cut out around that half of the pattern; this forms the parting of the boxes. The surface of the parting is well polished, some parting-sand thrown on, and the other half of the pattern set upon the first. Cores are now to be provided in those places where the upper box will not lift. We find that a core is to be made between the two forelegs, as indicated by the dotted line. Another core is needed on the face, from the nose to the ears; and a third core, joining the second at the ears, running down its neck. This will be all the cores needed, for the other parts of the pattern divide naturally. These cores are made of fresh sand, even if the other mould is made of old sand. Old sand will not stand the necessary moving of these cores. The cores are often moulded upon fine blotting or oiled tissue paper, if small; but if the cores are large, wire is to be buried in them. When the upper box is filled with sand, which is done after the cores are well finished and parting-sand put on, the upper box is lifted, one half of the pattern removed, and the flask closed again. The flask is now inverted, the lower box lifted, and the other half of the pattern removed. The same manipulation, in principle, is used in moulding a grooved pulley. By this mode of moulding, the cores are not removed. This is only practicable where the pattern can be and is divided, and where it is of light material. If the pattern is heavy, made of metal, and it cannot be divided, then the cores are to be drawn back from it as soon as the upper box has been lifted. There is no need of moving the cores further than is just necessary to have them out of the way for lifting the pattern. Good parting-sand is sufficient to separate cores so large as to take wire. Very small cores are best bedded upon paper; in pulling the paper, the core resting upon it will follow. As soon as the pattern is removed from the sand, the cores are again put in their places, and the boxes removed to the drying-stove for drying. It is a matter of precaution to fasten the cores, if they are once in their places, with hooks of thin iron wire, bent at one end, and pin the cores to the mould. There is less danger of injury happening to the mould, in handling the boxes, if the cores are secured in this manner. When the mould is properly dried, which may be done within twelve hours (though a longer time would be preferable), it is joined together, glands or screws put on, as the case may be, and cast. If the article is to be cast in bronze, brass, or any other metal besides iron, it is not blackened; but if it is to be cast in iron, it is blackened as before described. There will be no difficulty in casting the antlers to this pattern: the platform also is very simple. Moulders who are skilled in this kind of work, will mould two loose cores, one upon the other, but in most cases it is preferable to dry one part of the mould with its cores, and then put on the other cores; in both cases, however, it requires experience to handle such tender, brittle things, as small sand cores, often but one-eighth of an inch thick, and half an inch in area.

Ornamental compositions are screwed together, to form an ornament of many parts. Small ornaments are soldered together, or riveted and soldered. Solder for iron trinkets is a fluid composition of a little silver and gold. The soldering is performed by the blow-pipe. Solder for brass and bronze is the same, if the articles are to be gilded; in ordinary cases, brass or tin solder is used.

Brass ornaments are mostly cast hollow; this is not so much for reasons of economy, as with a view of making more perfect castings, and saving labour in chipping and chiselling. As no coal can be used to protect the metal against burning together with the sand, it is necessary to cool it as quick as possible, and in this way give it a smooth surface. The making of cores in these instances is often connected with considerable difficulties. The cores of complicated figures are composed of parts, that is, a corebox is made for one part of the core, and the parts cemented together to form the core. Iron castings are but seldom cast hollow, if small, that is, articles of less than six or eight inches extent; larger figures in iron are cast hollow, for if the body of hot iron is large, it will burn the sand, or melt together with it. Fluid iron, suitable for small castings, and the use of good fine sand, will make ornaments finer and sharper in expression than castings in any other metal. Horse-hair and cotton thread may be imitated to perfection. The wings of a fly with its microscopic nerves may be copied in iron; and green leaves stiffened so as to be applicable as patterns, may be cast in iron without difficulty.

Loam-Moulding.—This is the most ancient branch of moulding. In this department the moulder is his own pattern maker. He furnishes in most cases the pattern, and makes the mould also. In some instances a pattern, or parts of a pattern, are made of wood, and buried in the loam, but these cases do not happen frequently. The loam-moulder will furnish patterns with great ease, which cannot be made well or so cheaply in any other way. Any form of a pattern, or any casting of whatever kind, may be done in loam. In practice, loam-moulding is generally restricted to forms which cannot be cast conveniently in any other way. Loam-moulding is more expensive, generally speaking, than any other kind of moulding, except in cases of simple forms and heavy castings.

Every piece of loam-moulding is a regularly constructed edifice. No moulding in loam for a casting of importance, is commenced until a perfect plan of the whole operation from beginning to end is laid down. If no such plan is made, it may happen, and frequently does happen, to be impracticable to mould in the way commenced, whereby often the labour spent so far, is lost. The most important part of this branch of moulding, is the composition of the loam employed; it demands the strictest attention, and is varied according to the objects to be moulded, as loam suitable in one case will not answer in another. Fineness and porosity, and as little shrinkage in being dried as possible, are indispensable qualities. The mould must be dried hard, to resist the pressure of the fluid metal, which will otherwise break it or crumble it to dust, and spoil the casting. If loam is too close, or imporous, it will retain the gases developed by the heat of the metal, and cause either the metal to boil and make porous castings, or in the worst case cause explosion, and throw out the hot metal. If loam shrinks too much in drying, it will inevitably crack, make crevices into which the hot metal runs, and what is still worse, some parts of the facing of the mould will be pressed back, which causes uneven, rugged castings. The most important quality of loam is its porosity. The heat of the cast, and the presence of gas-generating material in every part of the mould, renders it necessary that the gases should escape through the substance of the mould, while it is impervious to the metal. There is no use in piercing holes by the pricker; if the loam is too strong, the cast will boil.

Moulding-loam is generally artificially composed of common brick-clay, and sharp-sand. Instead of the latter, old coarse foundry sand, or used core-sand, or burnt brick-powder, may be used to greater advantage. The quantity of sand to be mixed with the clay can only be known by experience. It is impossible to give receipts for compositions, because the quality of loam as well as that of the sand is variable, and differs in every instance. For heavy, thick castings, the loam is to be stronger than for small or thin castings.

Cow-hair obtained from tanneries is used to prevent the cracking of loam and make it porous. Mill-seeds, sawdust, horse-dung, hacked-hay or straw, are still more extensively used than cow-hair. Loam is to be worked diligently, to make its texture as uniform as possible. The matter to be mixed with it is to be distributed equally through the whole mass. Each part of the mould requires a different kind of loam: one for the facing of the mould, and another for the body; a stronger loam for brick-work, and a weaker one, with more straw or horse-dung, for a common mould. Parts of a mould which are almost surrounded by the pattern, and of course by the metal, are to be burned in a fire almost to a red heat, not only to expel water, but also to destroy everything which could generate gas, and to destroy every particle of vegetable and animal matter. This operation is necessary to be performed on all cores, and such parts of a mould as form the interior of it. The gases generated in a loam mould are of a complex nature; there are gases of water—steam—carbonic acid, carbonic oxide, and ammoniacal compositions which are not determined. The flame issuing from a loam mould, generally burns with a blue light, interspersed with greenish yellow streaks and specks.

Moulding of Simple Round Forms.—Articles of a round form, that is, a spheroid, or a segment of it, a cylinder and its auxiliaries, are moulded by means of a loam-board fastened to an iron spindle, which may be turned upon an imaginary axis, or the axis of the spindle. Wherever a loam-mould is built up, it must be always in the sweep of a crane, or it is to be built in that pit where it is finally to be cast. We will commence our illustration by the moulding of a soap-kettle in the pit. A soap-kettle—or soap-pan—is generally partly cylindrical, with a round bottom, broad brim, and a collar, for the wooden superstructure of planks to be set into it. All kettles are moulded and cast in an inverted position, as is shown in figure 21. It would be better for the quality of the cast if kettles could be cast bottom down, but this is almost impossible on account of the core. The moulding of a kettle is represented in figure 21.

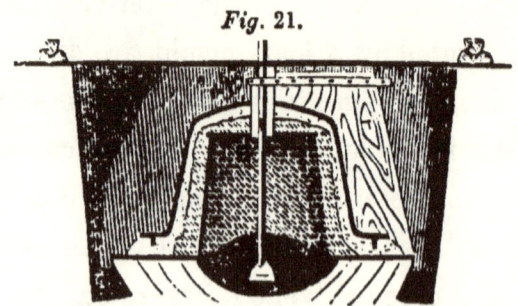

Fig. 21.

It is here performed, for want of a crane in the pit, on the very place where it is to be cast. A hole is dug in the floor of the foundry sufficiently deep to bury the whole mould, and wide enough to permit tne moulder to walk around the mould when he is at work. The first thing to be done is to cast a round plate or ring in open sand. This is to reach at least six inches into the kettle: that is, its smallest diameter is to be twelve inches smaller than the shorter or interior diameter of the kettle, and its largest diameter is to be from eight to twelve inches longer than the longest diameter of the pattern. This plate may be three-quarters of an inch or one inch thick. It is placed in a perfectly level position on the bottom of the pit, and raised by bricks to the height of six or eight inches from the bottom. In the centre of this ring-plate a pole or piece of cast iron is driven in the ground, and covered by sand to protect it against heat. In this pole a pan, or step, is cut for the pivot of the spindle to move in. A spindle of one and a half or two inches square wrought iron, having a round bearing at its upper end, and a steel point at its lower extremity, is now erected; resting below in the centre step, and above in a plank laid across the pit, borne down and held in its place by weights placed upon it at the extremities. This spindle is to stand perfectly plumb, being exactly in the centre of the foundation plate. To this spindle a moveable arm is screwed, made of two rods of flat iron, with many holes in it. At the spindle these two flat bars are bent in such a manner as to catch two corners of the spindle, and be immoveably fastened to it by two screws. In other cases a cast-iron forked arm is made with holes for screwing on the loam-board, and a spindle-box with pinching-screw for adjustment. To this arm the loam-board is screwed, which is an inch thick pine board, clear of knots. The loam-board is at first a skeleton of the interior of the kettle with the brim, and that slanting part beyond the brim, called the knee; if turned upon the axis of the spindle, it will describe the form and size of the interior of the kettle. In commencing the mould, a four inch brick wall is built upon the foundation plate, or platform, round, so as to leave two inches space between it and the loam-board. At the height of six inches below the corner of the bottom, a layer of one and a half inch iron bars is laid, and these are crossed by smaller bars, all walled-in in the brick work. Upon these bars the bricks forming the crown are founded. If the bottom is round, forms half a sphere, these iron bars are not needed; an arch may be sprung of bricks, which generally is strong enough to resist the pressure of the fluid metal. The moulder leaves a small opening around the spindle, serving the purpose of a drafthole for the fire which is to be kindled inside. This first brick wall is dried by a stone coal or charcoal fire, kindled inside below the mould. The loam-mortar for putting up the wall, is to be porous, out also strong; some horse-dung is generally mixed with it. It is composed mostly of sand, and the layers of mortar are from half an inch to one inch thick. The bricks used for this purpose are hard-burnt, light, but such as have not been melted, or burned too hard. Bricks are used in halves or bats. While the brick wall is drying, a thin layer of loam may be given to the mould, which here constitutes the core, in case the work is pressing; but if there is time, it is better to dry the bricks first. The loam may after this be increased to within a quarter of an inch to the loam-board, still keeping up the fire, and drying the core gradually. The last layer of loam is put on when the first loam-coating is nearly dry. It consists of finer and stronger loam. It is free from horse-dung, straw, or any other admixture, but is mixed with some cow hair. The surface is finished off by a smooth coating of wet fine loam, the redundancies being swept off by the loam-board, which has been washed and freed of all adherent loam and straw. As the surface gradually dries, it is painted over, by means of a paint-brush, with a mixture of charcoal-powder, clay, and water. This coating forms the parting between the core and the metal-thickness.

The loam-board with which the core has been formed is now removed, and is replaced by another board, called the thickness board. The edge of the thickness board describes the external form of the kettle, and merely touches the knee made by the first board. We see here the use of the knee: it serves not only for the more perfect closing of the mould below, but it is the standard mark of the loam-boards. Over the nearly dry core a layer of porous sandy loam is now spread, and made smooth by sweeping the thickness board around it. This layer of loam forms the exact pattern of the kettle as it will be after casting. When well dried, this layer of loam receives a blackwash as the core did, and is to be well dried. The spindle may now be removed, for there is no more use for it in this instance. Over the first foundation plate, or platform, is now laid another platform, whose interior diameter is equal to the diameter of the knee, so that this ring when laid down just fits, or is a little larger than the largest part of the core. Upon this platform another layer of loam of nearly two inches thick is laid, and smoothed over by hand. There is no need of a smooth surface. As long as the loam continues soft, the mould is kept under bars of iron bent in the shape of the bottom, or crown of the mould, and reaching down to the platform, to which they attach themselves by being bent under the platform. Two or three of such bars reach all over the kettle, others may be shorter and reach merely along the sides. These bars are laid over the soft loam, and then the mould is gradually dried. When nearly dry, iron hoops, which keep together the rods, are laid around the mould, and fastened to the rods by means of wire. The whole mould, iron and all, receives after this a good coating of straw loam, with horse-dung, the iron bars being partly covered with it. In this manner, iron and loam are combined and form one solid part of the mould. The structure of the mould is now completed, so far as the form is concerned. The whole is now thoroughly dried or baked by keeping up a constant fire in the interior of the mould. Fire may be applied on the outside also. In this instance, which is that of moulding a kettle with an open core, not so much fire is required as if the core was solid. In the latter case it requires a thorough burning; the core is then to be made red hot; but in this instance a good drying is sufficient to secure a safe cast. In twenty-four hours the mould will be found to be dry, and ready to be taken apart.

The taking apart of the mould is done by means of a crane, in case there is one; otherwise it is to be done by hand, by a pulley, or by some other means which are sufficient to lift the cape or cope. The first step to be taken is to lift, by means of a sharp crow-bar, the platform of the cope from the platform of the core, that is, to loosen the first from the latter part, after which it may be lifted and set upon a pair of timbers over the pit, or on any other convenient place where it is not exposed to moisture. When the cope is removed, the "thickness" is peeled off from the core, the draft-hole in the crown is closed by a brick-bat and plastered over with loam. The hole in the centre of the cope is also filled up to within two inches, all the damages on the mould repaired, and these patches dried. After this the mould receives a blackwashing, and is then finally dried once more.

Blackwash.—The blackwash for parting consists chiefly of charcoal-powder, and a little clay. This is almost entirely lost in taking the mould apart, and the remainder is lost in sleeking the mould by the finishing clay-wash. Blackwash is always on hand in the foundry; it is contained in the blackwash tubs, of which there is one for parting and one for finishing. The latter is composed of finely ground plumbago, often mixed with a little charcoal, the whole diluted with horse-dung water, or a solution of the soluble parts of horse-dung. This blackwash is frequently mixed with pease-meal or other meal, glue, and extracts from the refuse of tanneries; but all these latter compositions are more or less too close, and cause a dull surface to the cast. The first is the best, if applied not too much diluted.

The sleek-washing as well as blackwashing is to be done with proper caution, so as not to injure the sharp outlines of the mould; it is better if the first of these two operations can be dispensed with, and the mould finished just as the loam-board left it. This latter can be done in following the plan to be described in cylinder forming, which is also applicable in this case; that is, working without thicknesses. When the parts of the mould are properly dried, it is put together again; but before this is done, a hole of two inches round is cut in the brim of the cope, to connect the gates with, for casting. The cope is to rest firmly upon the core, that is, it is to be put exactly in that position in which it was before, and shut tightly at the knee. A pipe is now laid below the foundation of the mould, which pipe is to be carried through the sand which is subsequently rammed in, to conduct the gas from the interior of the core to the surface. This pipe may be either an iron pipe, or may be of baked clay, or it may be a space left in the sand. The latter is objectionable, because it may fill up, and cause an explosion by stopping the escape of the gas. The mould is now rammed in with sand, which is done by iron stampers with strokes in rapid succession. This operation is performed by at least three hands at once, to break the vibrations caused by stamping, and prevent injury to the mould in consequence. With the ramming-in of the mould, the gate for the reception of the metal is to be provided for. This we contemplate to be in the lower part of the mould; it is frequently done from the top, but the latter mode is not quite safe, and never makes as sound castings as the way proposed here. The gate may be formed by a wooden pattern or pin, as in green-sand moulding, but this is at best a very doubtful operation in its consequences; for the gate will be a long one in all cases, and in pulling out the pin, sand may drop in the gate and stop it up altogether, or drop s much sand as to injure the casting. The best plan is to have pipes ready made of burned loam; such pipes may be conical, and tapered so that the smaller end of one pipe will fit in the larger end of another. In this way any length of gate may be formed, perfectly secure against damages from stamping. On the top of the mould a flow-gate is set, which may be also formed of one of burned loam-pipes, or it may be moulded in the sand. The first plan, however, is preferable. The whole space around the mould is in this way filled up with sand, and tightly rammed. Over this sand, that is, over the mould covered by the sand, pieces of pig-iron or other heavy pieces of iron are laid, or iron beams tied down by screws which reach to the platform, and are fastened to the latter, to prevent the least motion of the mould upwards, for such a motion would spoil the mould. Before casting, the flow-gate is covered with a dry ball of loam, to be removed when the fluid metal shows itself in the gate and the mould is filled with iron. Over that channel or pipe, communicating with the interior of the core, a handful of dry wood shavings, or dry straw, is laid, and kindled as soon as the hot metal is being poured in.

The stopping up of the flow-gate is a necessary operation, and the flow-gate itself also is necessary in all cases of large castings. The flow-gate is very useful, because it is always put on the highest point, or at a point to which most of the light impurities which float on the melted metal are very apt to flow. If the flow-gate is placed in such a situation, the impurities will naturally flow into it. For these reasons the flow-gate is always made wider than the cast-gate. The stopping of the flow-gate until the metal appears, is an operation equally important. If the flow-gate, or any other aperture to the interior of the mould, is open, the gases or hot air will rush to the opening with a force equal to the space of the mould and the amount of hot metal to be poured into it. This rush of air is very apt to tear loose some loam or sand of the mould, or even break the mould. By stopping the openings, a certain amount of confined gas finds its way through the sand or loam of the mould, and opens the pores of the mould. This stopping up of the air channels is the safest way of preventing explosions and making good castings. In case there is no flow-gate to a mould, and only a cast-gate, the latter is to be kept full all the time during which metal is poured in. If there is an interruption of the flow, and the rush of air finds its way through the cast-gate, it is very apt to cool the metal, tear some sand loose, and by that means stop up the gate, or even break the mould. Such accidents happen frequently, and are the common causes of failure in founding. When castings are made by a single cast-gate, it is advisable to make a reservoir for the fluid metal at the top, that is, to make the mouth of the gate very wide, and skim the metal well to prevent the flowing in of any impurities. In moulding hollow-ware, the wedge-shaped gits are made partly for causing an easy separation of the git from the cast, but chiefly to have a git of large capacity and small opening, to be kept full while casting.

Gas Pipes.—The air pipes leading from the core of a heavy casting ought to be made of iron, for these pipes have an important office to perform. In case such a pipe is stopped up, an explosion is almost inevitable. The atmospheric air confined in the hollow space of the core, and that air contained in the pores of the sand, mixed with the carbonic oxide gas generated of the vegetable or animal matter in the mould, will form an explosive mixture of the most dangerous kind, and will destroy any mould if it explodes. The mouth of the air pipe may be covered with burning shavings, but it is advisable first to lay over the mouth of it a piece of wire-gauze, to prevent the falling in of any dirt or fire. If there is fire in the pipe before the air is moving, that is, before there is any metal in the mould, an explosion will take place.

Removing of the Core.—As soon as the casting is done, the mould is dug up, and a portion of the core removed before the cast is entirely cooled. Cylindrical castings are liable to be split by the core, if the core is too strong. For these reasons the core is made chiefly of sand, and only enough of clay is used to keep it together. Brick cores are preferable to loam cores, if put together with sand and thick joints, because bricks offer some resistance to the fluid metal by their mass, and are easily moved by a strong power, such as metal in the act of contraction. This is one of the evils attending iron core pipes. If there is no hay-rope or a thick layer of sand around a core-iron, the casting will split upon the core before it is cool, and before it can be prevented. In all cases it is advisable to remove the core as soon as possible, and if it cannot be taken out altogether, then remove at least a part of it, that is, cut it in some place so as to afford room for the contracting cast.

Moulding without Thickness.—As an illustration of moulding in loam without thickness, which is certainly the most advantageous plan of loam-moulding, we will describe the moulding of a cylinder

MOULDING.

The operation is similar in all cases: whether for a steam-engine, a blast-machine, or a cylinder for any other purpose; for illustration, however, we prefer that of a steam-engine, as the most complicated. In cases of narrow cylinders it is preferred to have the core fixed, and move the cope, particularly where the latter is to be divided. Dividing the cope ought to be avoided, if possible, for it is almost impossible to make a correct casting in such a mould. We will take a case for illustration where core and cope are each in one piece, and the latter stationary, that is, moulded in that place where the cylinder is to be cast. In this instance the mould for the cope is put in the pit, the same as the mould of the pan, above described, and founded the same way upon a platform. It is not advisable to make the cope above ground, even if there is a crane strong enough to carry it to the pit. In a mould like this, a crevice may open in transporting it, and give access to hot metal, which may frustrate the purpose for which the mould has been made. In figure 22, the moulding of a short cylinder is represented, such as is now used in steam-engines to turn the screw propellers of steamboats. A pattern of the steam-ways is made in wood, solid, as represented in figure 23, which figure shows a side elevation, and a view from above. This block has the length of the cylinder between its flanges, and in

Fig. 22.

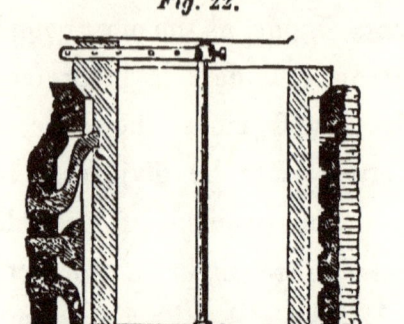

Fig. 23.

case there are any mouldings in the flange which run across the steam-ways, they are to be moulded in the wood. The three core-prints are of considerable length, because the cores find here their chief support. The middle core finds another support in the opening for the exhaust pipe, as shown in figure 23. One side of the pattern is hollow and cylindrical, fitting the exterior diameter of the cylinder, or the sweep of the loam-board. Having laid the platform, erected the spindle, and screwed in the loam-board—which is almost a straight board, with the exception of the two knees, one above and one below, and the moulding or hoops around the cylinder—the brick enclosure is laid, leaving from two to two and a half inches space for loam. The pattern of the steam-ways is fastened, just touching the loam-board in its travel upon its axis, and walled in, giving it a loam coating where the bricks touch it. After the brick wall is nearly dry, a coating of loam is given; this loam may be pretty strong, and mixed with hay, for the pressure upon it will be great, and if the loam gives way to this pressure, the cylinder will be defaced. This coating is superficially dried, and another coat of hair-loam laid on, which is to reach very near the loam-board, so that the last coating is but a little thicker than a clay-wash. In drying the previous loam coats, and making the coats thin, an almost perfectly smooth surface of the mould may be obtained. It will be as round and straight as a turned and polished iron cylinder, and of course the casting will be similar to the moulding. Clay which shrinks a great deal, or is plastered on in too heavy coats, causes uneven and rugged surfaces in the mould, which is the case also if the ground is not dry before the last washing is performed. A good facing is as smooth, sharp, and distinct in its outlines as a well made pattern of wood. The blackwashing is here to be the very last operation, and to be well performed, and when dry must be polished by a large sleeker fitting the circle of the cylinder. Before the blackwashing of the cylinder is performed, however, the steam-ways are moulded; while the cylinder is under the influence of the fire. The pattern of the steam-ways is covered by hair-loam, leaving the core-prints projecting, so as to afford access to the interior through the holes left by the core-prints. The pattern receives two or three coatings of loam, sufficient to make the loam at least two inches thick. Close to the brick, or as far off as the square of the pattern goes, a groove is cut in the loam, around the pattern, indicated by the dotted line, figure 23. This groove cuts the loam nearly through, so that the mould may be separated at this mark. The mould around the steam-ways pattern is provided with iron, bent around it, and also irons bent around the cylinder. These irons meet at the joint or parting, and are secured in their places by wire and loam, the ends of the irons at the parting terminating in hooks. Fastening a mould in this way by iron straps is convenient and advantageous, and answers every purpose, if the mould is made strong enough. If a cope is made too weak because of the iron straps, the above is a bad fastening, and the cause of failures or imperfect castings. Fastening a mould with iron is expensive, and where it can be avoided it is advisable so to do. In this instance it can be avoided, and the mould may be made serviceable without iron fastenings. When the steam-ways pattern is removed, and the mould ready to be closed again, it is simply closed and secured by brick-work, which latter is commenced at the bottom. While the brick-work is progressing, the cope of the steam-ways is secured temporarily by some wire fastened around the cylinder. The brick-work forming the enclosure to this part of the mould is dried by external fire, or the united heat of the fire inside and outside of the mould.

The cores, forming the steam-ways, must be strong and porous. They are to be as long as the hollow they are to form in the casting, to which is to be added the length of the core-prints. Cores of this description are generally moulded in a wooden core-box; but this plan is not to be recommended, for wood will twist and warp, particularly where it is wet on one side and charring hot on the other, as is the case in this instance. The best plan of mak ing the cores, is to make a wood pattern of a core-box, and cast it at once in iron and in open sand. In such an iron box, a good and correct core may be made without much labour. The cores for the steam-ways are made of strong loam, and provided with several core-irons, which are rods of quarter or half inch square iron, bent in the curves of the core. The core-irons are dipped in strong clay-water before they are buried in the core.

Besides the core-irons, strings of hemp, cotton, or straw, are laid in the core, which burn out in drying and form channels for the escape of air. A great many of these strings may be used, but they must be thin, so as to arrest the fluid iron, in case any of it finds access to the interior of the core. The core-loam may contain cow-hair if necessary, but this is a matter which depends entirely on the quality of the loam of which the core is made. The cores, after being moulded, are heated to redness in a coal fire, with liberal access of air, to expel every trace of water, vegetable and animal matter, and carbon. When well burned, the cores receive a good black-washing of black-lead and clay, as little as possible of the latter. These cores are the very last to be put in the mould.

Core for the Cylinder.—While the cope of the

cylinder is progressing, the core for it is moulded somewhere near it, on the floor of the foundry, but within the sweep of the crane. The core is founded upon an iron platform, which has its snugs inside, and its diameter is six inches less than the diameter of the interior of the cope. The platform of the core is to rest upon the platform of the cope. The core is

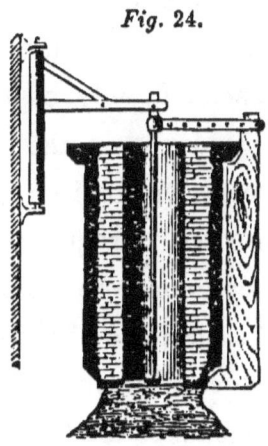

Fig. 24.

simply built of bricks, finished in loam, blackened and polished, and is then ready to be set in its cope. The core has two knees, one below and one above, which are at an angle of 45° These two knees are necessary to keep the core in its position. In case the metal is liable to porosity, which is frequently the case with some of the anthracite iron, and generally so with charcoal iron, it is necessary to prolong the mould of the cylinder, above its flange, as shown in the drawing, figure 24, into which the sullage rises. In cast iron which does not form holes, or raise any sullage, this precaution is not required. Upon the sullage piece, or in want of that, upon the upper edge of the cylinder, the flow-gates are set, of which there are to be at least two or three, and more if the iron is doubtful and the diameter of the cylinder large. Before the core is put in its place, two rests for the steam-way cores are cut into it. The steam-way cores are suspended only at their two ends, and liable to be lifted out at the centre core. A deep rest in that core, or an iron fastening which passes through that core, is required to secure it in its place, when the cylinder core is set and well secured, resting upon the platform of the cope, where it is secured by iron wedges. For these reasons the knees of the mould may be made to catch before the platform plates touch one another, and the space left between them can be filled up by iron wedges or scraps. The cores of the steam-ways, when put in, are well secured to the core, and then the cope of the steam-ways put in its place. The cores are after this secured in the openings left by the core-prints of the pattern, and well stopped up by moist loam, which is to be dried. In many cases, that straight part of the steam-ways cope where the cores pass through, is covered by an iron plate, coated with loam, and the core-irons fastened to this plate. This caution is unnecessary, as the projecting cores can be well secured by dry brick-bats. The mode of fastening, however, depends very much on the size and form of the steam-box, and the form of the cores.

The burying of the mould and ramming-in of the sand is done in the usual way, but here the space below the mould is filled with sand and well secured, to prevent the hot metal entering below the core, in case the lower knee does not fit tightly, which in this case is always doubtful, and cannot be secured beforehand. The interior of the core is also filled with sand, if there is any doubt of its being strong enough and tight. It is better when there is no sand in the core, at least but very little in the bottom of it. The opening of the core at the top is covered by an iron plate which is well secured, leaving but a small opening for the escape of the gases; which opening is, as in any other instance, covered by a piece of wire gauze and burning straw. The whole mould is covered, as well as the core-plate, with a load of iron or screws, to prevent any motion of the core or cope by the static pressure of the fluid metal, for the least lifting will inevitably destroy the cast. The cast-gate is at the lower flange, and the metal is to rise gradually from below.

The cores of the steam-ways are often of such forms as not so easily to be secured in their places, which is particularly the case with the middle, or exhaust core. In this, the assumed case, there is no difficulty; for we have two strong core-prints, and the core cannot be large, as the steam-chest is but small. If a core-print can be given on each side of the chest, there will be no difficulty at all, for then the core has three points to rest upon, and can be made safe enough. If the other two cores are strong enough to take strong core-irons, there is no danger of their failing. Where such advantages cannot be had, and where the cores are in danger of being lifted off their seats, it is necessary to secure the cores by chaplets, which are put between the cores and the cope of the steam-ways, for there are none applicable to the core of the cylinder.

The use of chaplets in the steam-ways cannot be recommended, if it can be avoided by any means. The chaplets must be strong and of good wrought iron, or the fluid iron will melt or dissolve the chaplets, and the effect is worse than if they had not been used; for the moulder depended upon a support which failed, and would have done better without supports. If chaplets are not made of good and very pure wrought iron, they are liable to melt, or are dissolved in the mass of cast iron. The greater the amount of the latter and the longer it keeps fluid, and the hotter it is, the greater is the danger of the chaplet being destroyed. Impure iron, or iron which contains much cinder, or thick scales of hammer-slag, is apt to produce holes in the casting, for the oxygen of the scales, or cinder, will combine with the carbon of the cast iron and form carbonic oxide, which cannot escape, as it is in the interior of the casting, and the iron next to the mould is generally chilled before such gas appears.

General Remarks on Loam-Moulding.—Precautions which are to be taken in loam-moulding in general, are to be particularly observed in moulding steam cylinders, for here the object is to make a smooth, well finished casting, and one of compact sound metal, free of pores or holes. To accomplish this, particular attention must be given to the following requisites: A strong but still a porous loam; drying in coats; a well smoothed facing before the blackening is put on; well burnt cores for the steam-ways, and the air-holes in these so small and so arranged, as to prevent any possible entrance of hot iron into these air channels; the absence of all chaplets if possible: and every part of the mould well dried. The bearing down of the mould, and the stamping in, are operations which are in all cases the same.

If there are any square or unusual forms on a cylinder, as, for example, if one or both flanges are square, or if there are extra steam-ways, or ornaments, all such forms are made in wood or in metal (the latter is preferable), buried in the mould, and removed before the finishing of the mould.

Irregular Forms.—Where forms are to be moulded which do not permit the use of the spindle, a loam-mould is made either by hand, or over a wood pattern. There are also cases where both instances happen in one mould. We will illustrate this by giving an instance of the first and an instance of the latter case. In figure 25, a bent pipe is represented, which cannot well be moulded in sand, and for which

Fig. 25.

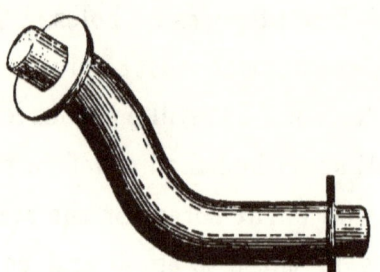

a loam-core is to be made in every instance. It may be moulded in sand or in loam. This pipe forming a knee, is bent in such a way as to make the moulding over a wood pattern and in sand almost out of the question. The first step taken is to make a drawing of the actual size of the object upon a board, and in drawing two or three sections of it, giving sufficient length for the core-prints. This board is given to the blacksmith, and one or more bars of iron bent in the shape of the core, and these bars united to form the core-iron. If the pipe is more than eight inches in diameter, these bars are to be laid around small rings, forming in this way an open channel in the centre of the core. These iron bars are covered with hay-rope as usual, and then by loam, which latter is laid on by hand, referring repeatedly to the drawing. The last loam coating is thin and well smoothed off, before the parting-blackwash is given. In such cases as this, it is all important to have the flanges at the right distance and in correct angles; and as such castings generally are designed to fill a space or form a connexion between two pipes, it is necessary to form a skeleton pipe of two boards, of which each fits to the flange of the corresponding pipe. Such a skeleton is easily formed by nailing boards together in that place where the pipe is to be. Figure 26 will show how it is performed. The boards are fitted and nailed together, stayed by some lath, and the place of the flanges marked by scribing around them. Over

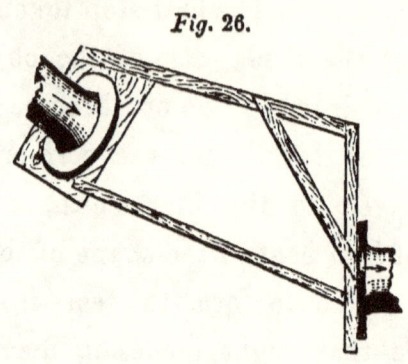

Fig. 26.

this another skeleton of boards is made, so as to have the dimensions of the pipe inside which are here outside, with the addition of one-eighth of an inch for each foot in the length of the pipe, for shrinkage. In this latter skeleton the inner diameter of the pipes is marked and cut out, the newly made core laid in this board skeleton, in the exact position in which the new pipe is to be attached to the other pipes. The core is fastened in this position to the skeleton, and the "thickness," which of course includes the flanges, is laid on the core, and gently dried. When the thickness is so far dried as to be secure against warping, it is removed from the skeleton boards, dried, blackened, and the cope put on. If the pipe is heavy the cope is to be fastened with iron, taking care to have the parting free. Moulds for light pipes may be secured by a succession of wire fastenings which are laid at certain distances around the cope. The parting of the cope is done as usual, by cutting two grooves along the pipe in such a direction as to divide the cope into two halves, but so that each half may be lifted off the core. If the flanges or the thickness break off in removing the cope there is no harm done, if the core is not damaged in this operation. After the usual finish of the facing, the mould may be put together, and rammed in sand as usual. In this case the core cannot be kept in its place without chaplets, and a liberal number of them is to be distributed between the core and the cope. This pipe is rammed in and cast in the usual manner.

When the object to be moulded presents more complicated forms than the one represented, the experience of the moulder must be his guide in forming the plan of the mould. Analogous processes are here everywhere, but it is the sagacity of the moulder which gives to the most complicated forms tangibility, which analyzes a pattern, and finds a mode of execution in cases where success at first sight appears to be impossible. If the form of a pattern does not happen to be divisible into two parts, or permit a mould of two parts, there is no objection to dividing it into three, four, and more parts, but it is a rule to make as few partings as possible. In every mould, it is to be a standard rule to provide liberally for the escape of the gases. If forms are to be moulded which require more than two platforms, there is no objection to taking as many as may secure the greatest advantage and security to the mould.

Oval Forms.—Oval, curved, or triangular forms must be traced by corresponding platform-plates, for no application of the spindle is possible in these cases. For example, to mould an oval bathing-tub, without a pattern, a foundation plate in the form of the upper side of the tub is cast in open sand. There is no need of its being solid— it may be an oval ring. Figure 27 represents the

Fig. 27.

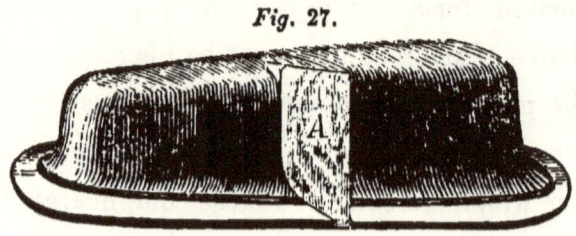

moulding of such a tub. The loam-board A is guided by hand around the platform, and if kept in close contact with the edge of the plate, there is no difficulty in obtaining a correct mould. If there are any projections, or departures from the regular form, they are made by hand. Curved forms are made in a way similar to the above. A core, or a mould to an elbow pipe, is moulded on a platform which has the form of the curved pipe, as shown in figure 28.

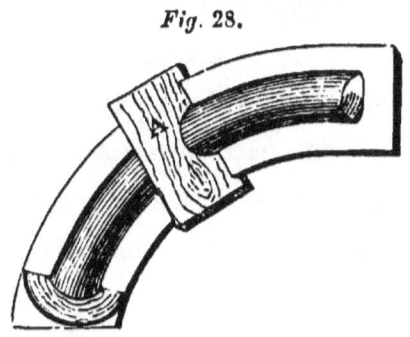

Fig. 28.

The loam-board A can make only the current part of the mould, also a mouth or bell-shaped widening; but if there are any flanges, for these a wood pattern is to be made. In this instance two halves of a pipe-core are made; and these joined by moist loam and wire. In most instances of this kind a wood pattern of the object is made, and this moulded in sand; but as the core of such forms cannot well be moulded in sand, it is made in loam and applied in the usual way. Square forms of objects which are to be moulded without patterns, are made in a similar manner as those of an oval or irregular form; such moulds, however, require more strength than the moulds of round forms, for the pressure of the fluid metal upon a plain surface, tends to drive the core and mould apart, with more energy than it does in round forms. To guard against this pressure in flat or straight forms, is an object which requires some judgment on the part of the moulder.

If complicated forms are to be moulded, the best plan always is, first to make a pattern in wood of the object. Even if the pattern is not used in moulding directly, it is of great service to the moulder, in having a form to imitate, which is more plastic to his mind than a mere drawing. All heavy and complicated castings, such as heavy bed-plates for steam engines, housings, and rollers for iron works, are moulded in loam, if good work is expected. The heat and pressure of a mass of hot iron like that poured into the mould for the bedplate for the engines of the Collins Atlantic steamers, being forty tons or more, will destroy any sand mould, no matter how carefully made. Complicated forms of this kind are partly made to drawings and partly over wood or metallic patterns. We will illustrate this subject by an instance which is not complicated, but sufficiently so to show the principle upon which a mould of this kind is constructed. In figure 29 a screw-propeller is shown, such as are now frequently used to propel steamboats. These propellers are cast in iron, copper, brass, or bronze; this, however, does not cause an essential difference to be made in constructing the mould. The four wings of this

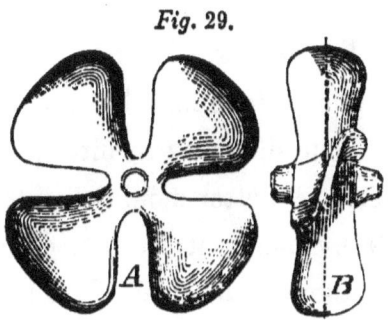

Fig. 29.

pattern are twised as shown in B. It is advisable to make a wood pattern of this propeller, dividing it at the dotted line in B into two halves. An experienced moulder will prefer to make the mould by hand, but generally the pattern is buried in the loam, and kept there until the mould is nearly dry. There is little difficulty in moulding this object in the latter way. As the pattern is divided, the one half is moulded upon an iron platform, the larger spaces filled by brick, and over these the usual coating of loam. The four wings of the pattern are fastened by wood-screws to the nave, which may be drawn and the pattern removed in parts; this forms the lower part of the mould. The other half of the pattern is moulded in parts, upon quadrant plates, with its dividing side downwards. The mould of this half is taken apart, each quarter resting upon its quadrant platform. These four quarters are set upon the first half of the mould which is whole, and has a solid platform. The edges of the four wings or paddles are generally sharpened out, so that there is little difficulty in hitting the thickness of the paddles. A better mould than that described may be made by hand; it is then divided into two halves as the above, but it affords a better opportunity of having the facings of the mould correct and uniform in texture. Many screw-propellers are moulded by dividing the pattern at the nave, and making a cope over each paddle, which is fitted and fastened to the cope of the nave. The first way of moulding is preferable to the latter; it is perfectly safe, and makes a more correct and smooth casting.

Moulding of Bronze Ornaments.—The art of casting bronze statues has been traced to remote antiquity, and, to all appearance, the ancients were more skilful than the moderns in this art. Bronze statues were so plentiful in Greece at the time of Alexander the Great, that Pliny calls them the mob of Alexander. It is recorded that the Romans found 3000 bronze statues in Athens, and as many in Rhodes. The Temple of Solomon was adorned with heavy and richly ornamented bronze castings. The pillars of Jachin and Boaz at the portal were of bronze; the molten sea of the priests to wash in, was cast of bronze, and the metal basins at the entrance were of the same metal. The world-renowned Colossus of Rhodes was a bronze statue of 130 feet high; it was broken by an earthquake fifty-six years after its erection, and its remains lay scattered over the ground for nearly nine hundred years, when they were sold by a king to a Jew, who carried at that time 360 tons of metal away. More recently, in the middle ages, bronze was extensively used for doors and gates of churches and cities. The doors at the Battisterio in Florence were of such exquisite workmanship, that Michael Angelo, the great architect of St. Peter's at Rome, declared that these gates were worthy to be the gates of heaven. More recently, in our own times, this beautiful art has been degraded to the manufacture of implements of war, and in other cases to celebrate the memory of military heroes—an application no better than the other. The ancients were not acquainted with a definite alloy, to make their bronze castings of. Their mixtures were accidental; but we will speak of this hereafter.

Moulding of Statues.—The mode of forming the moulds for bronze castings of large size, as statues and bas-reliefs, was never reduced to a systematic art. There is satisfactory evidence to show that the knowledge of this art lay dormant for centuries. The ancient Greeks were the most skilful in the execution of statues of this kind, not only so far as form is concerned, but also in their preparation of the moulds and the casting of the statue. Their plan of making a mould, was to make a skeleton of plastic clay, which was to form the core. This skeleton was kept wet—just as the sculptors of the present day mould a figure in clay—and made into an exact mould of the figure to be produced. Over this wet clay pattern the cope was made, and so far dried as to admit of removal, after which core and cope were finally dried and burned. The space resulting from the shrinking of the core, formed here the thickness for the metal. The way in which such a mould was made is an evidence of the high skill of the artists of that time; for in case the casting fails, all the labour of the artist and the moulder is lost, for pattern and mould are destroyed at each cast. It requires great experience and skill to succeed in this mode of casting statues and larger ornaments.

French Mode of Moulding Statues.—A more safe, but very expensive plan of making moulds, was practised in the seventeenth and eighteenth centuries. The pattern for larger statues was made of plaster of paris, instead of clay, because the latter shrinks a great deal in large masses. This plaster was laid on and fastened to a skeleton of iron. Over this pattern, which might be either an original or a pattern at hand, a cast of plaster is made, and this plaster mould divided so as to remove it conveniently. Over parts of this plaster mould coats of wax are laid, which form the "thickness." The wax is a compound of six parts of wax and one of white pitch, with which a little tallow or oil is mixed. The plaster mould receives a film of oil before the wax is put on, and the first coating of wax is laid on warm by means of a paint brush. A skeleton of iron bars is now made, composed of heavy and small iron, also iron wire and wire gauze, having, as near as possible, the form of the object to be cast. The segments of wax are fastened to this iron skeleton, and finally the whole surrounded by the plaster cope. Into this hollow mould, which is composed of the cope of plaster, a thickness of wax, and an iron skeleton in the interior, the cement forming the core is cast. This cement is composed of two parts of plaster of paris, and one of brick-dust, or ground bricks, cast through an opening made in a convenient place as high as possible on the mould. When this core is hardened, which takes but a short time, the plaster cope is removed, the damages in the wax mould repaired, and a number of small gits for conducting the metal, and other gates for letting out the gases, are fastened around the figure. These gates are made of wax, from half an inch to one inch thick, and fastened to the figure in such places where the least injury will be done. None are to be on the face, hands, or other delicate parts. Small wire is used to keep these gates in their places. The final cope is then made in the usual way of sand-loam, mixed with cow-hair, or horse-dung. The first coating on the wax figure, however, consists of finely ground brick-dust, mixed with the white of egg or glue, forming a kind of paint. This is painted twenty times and more over the pattern. After this first coat follows a coating of hair-loam, and finally horse-dung loam. This loam-cope is to be provided with iron fastenings, and at last receives a brick enclosure, which is also secured by iron binders. Below and around this mould fire-places are erected, which are so distributed as, when fire is made in them, to make the mould uniformly warm outside and inside, and heat it to an almost red heat. The wax forming the thickness is the first that flows out, and leaves a space in the mould of the same thickness as the cast is to be. The quantity of metal needed to fill the mould is exactly that space occupied by the wax. This process of moulding is complicated, but it is safe and insures good castings. It has the advantage over the Grecian mode, that the original pattern, the plaster cope, is never lost.

By skill and dexterity the artist may shorten the above process. One way is to build the plaster cope directly over the iron skeleton for the core, cast the mould full of core-cement, remove the plaster cope, and shave the "thickness" off the core. Then put the plaster cope again around this core, and cast the thickness space full of wax. Over this wax cast, the loam cope is made, as described above.

At the present time there is no settled system in the casting of bronze statues: the artists follow their own individual inclinations and experience. In many instances cores are built up first, covered by hand with loam, and burned; then the wax is put on, and the pattern made upon the core; over this pattern the loam cope is moulded, the wax melted out, and the mould filled with metal in the usual way. In this way the pattern is lost. In other cases they make a core as above, cover it by wax plates made in the plaster mould, and proceed as described before. All the difference from that described in the past pages is here the making of the core, which, if made in the latter way, is more perfect, and more certain to secure success.

Iron Statues require more metal than bronze statues, and also strongly burnt moulds. Here the core is built up first, and the "thickness" laid on in fine clay. The pattern is made by the sculptor upon the core. The cope is made and divided as in common loam-moulding, the thickness removed, and the mould put together with that caution required to make the operation successful. The pattern of course is lost, and if the casting fails it is to be made anew. A mould over a pattern at hand, may be made over that pattern, but the core is to be made by hand. In all cases core as well as cope are to be well provided with iron stays, and chaplets, and are to be perfectly dry. If such cautions are taken, there will be no failure in casting.

Bas-reliefs.—Flat bronze castings, as ornamented pannels, facings, and single ornaments, are cast in the usual way in iron flasks, in new sand, and dried.

If the patterns are too complicated, or underworked, so as to make many cores necessary, the facing of the mould is made in fine strong sand, entirely composed of cores, and over these cores, as a parting, the whole of the cores are covered with common moulding sand and dried all together. The parting between the cores and the sand is made by common parting-sand. To avoid the division of the mould, the patterns are frequently cut in such places and directions as to remove the pattern in parts. This latter mode of moulding, because it is the cheapest, is practised in the manufacture of articles which are in common use.

Moulding of Bells.—Small bells are generally moulded in sand, from a metal or wood pattern, and the sand mould is dried in a stove, as before described. We shall give no description of the manufacture of small bells, to which class bells of from one hundred to two hundred pounds' weight belong, but confine ourselves to a description of the moulding of the larger kinds. The most important part of this art, is the construction or the form of the bell. Another equally interesting is the composition of bell-metal. In this place we shall only speak of the moulding of a large bell. In figure 80, a mould is represented as it is sunk in the pit for casting. There is no essential difference between moulding a bell and a cast-iron kettle. The core is built in brick upon an iron platform, which is to

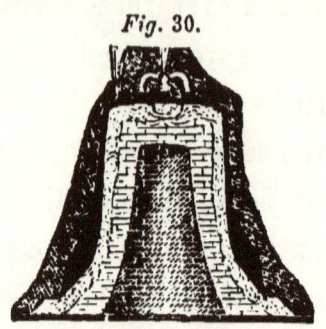

Fig. 30.

have snugs, in case the mould is made above ground This brick core is covered with three-fourths of an inch or one inch thick of hair-loam, and the last surface-washing is given by a finely ground composition of clay and brick-dust. This latter is mixed with an extract of horse-dung, to which is added a little sal-ammonia. Upon the core the "thickness" is laid in loam-sand, but the thickness is again washed with fine clay to give it a smooth surface. Ornaments which have been previously moulded, either in wax, wood, or metal, are now pasted on by means of wax, glue, or any other kind of cement. If the ornaments are of such a nature as to prevent the lifting of the cope without them—for the cope cannot be divided—the ornaments are fastened to the thickness by tallow, or a mixture of tallow and wax. A little heat given to the mould will melt the tallow, after which the ornaments adhere to the cope, from which they may be removed when the cope is lifted off the core. The thickness is to be well polished; and, as no coal can be used for parting, the whole is slightly dusted over with wood-ashes. The parting between the core and the thickness is also made with ashes. The cope is laid on at first by means of a paint-brush, the paint consisting of clay and ground bricks, made thin by horse-dung water. This coating is to be thin and fine; upon it hair-loam, and finally straw-loam is laid.

The crown of the bell is moulded over a wood pattern, after the spindle is removed. The iron or steel staple for the hammer is set in the core, into the hollow left by the spindle. It projects into the thickness, so as to be cast into the metal. The facing of the mould ought to be finished when the cope is lifted off. Small defects may occur, and are, if not very large, left as they are; the excess of metal in those places is chiselled off after the bell is cast. All that can be done in polishing the facing of the mould is to give it a uniform dusting of ashes. When the mould is perfectly dry, it is put together for casting. The core may be filled with sand, if preferred, but there is no harm done if it is left open; for bell-metal does not generate much gas, and there is no danger of an explosion. The cope is in some measure secured by iron, but its chief security is in the strong, well rammed sand of the pit. The cast-gate is on the top of the bell, either on the crown, or, if the latter is ornamented, on one side of it. Flow-gates are of no use here, the metal is to be clean before it enters the mould: there is no danger of sullage.

Moulds consisting partly of loam or sand, and partly of metal, are in frequent use in iron foundries. Small car-wheels, boshes for cart-wheels, and car-wheels for mining establishments, receive their bore by being cast over an iron or steel core. Such a core-iron is a little tapered, to admit of its being freed from the casting by a smart stroke of the hammer. The casting is never left to cool entirely before the core is removed. It is generally removed when the casting is hot, but so far cooled as to resist the drawing out of the core-iron.

Chilled railroad-car wheels are another article where iron is employed as a part of the mould. The cast and chilled railroad wheels now in general use, are cast in a mould composed of green sand and iron. In figure 31 is shown a mould in which a chilled wheel is cast. It consists of three boxes. The lower is a box of common round form, merely to hold the sand and give support to the centre core

Fig. 31.

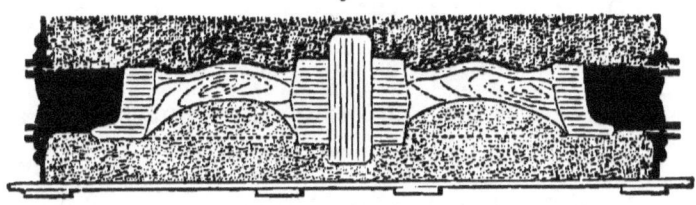

and the middle box. The upper box is of a similar form, also round. The middle box is a solid ring cast of strong gray or mottled iron, and bored out upon a turning lathe, giving it the reverse of the exact form of the rim of the wheel. This middle box ought to be at least as heavy as the wheel is to be after casting, and it is preferable if it has two or three times that weight. All the three boxes are joined by ears and pins as usual, and the latter ought to fit well without being too tight. The chief difficulty in casting these chilled wheels is to make the cast of a uniform strain to prevent the wheels from breaking. Wheels with spokes or arms are very liable to this evil, and are to be cast with their hubs divided into three or more segments, which are afterwards banded by wrought-iron tires before fastening them to the car axles. At present, most of these wheels are cast with corrugated discs or plates; in this way the hub may be cast solid, and the wheel is not so liable to be subjected to an unequal strain in the metal as when cast with spokes. In such plate-wheels the whole space between the rim and the hub is filled by metal, which, however, in most cases is not more than three-quarters of an inch or one inch thick. The rim of a good wheel is to be as hard as hardened steel at its periphery, but soft and gray in its central parts. The first requisite is more safely attained by having a heavy chill; but if the chill is too heavy, the inner parts are apt to suffer the cooling qualities of the chill. Success in this branch of founding depends very much on the quality of the iron of which the wheels are cast; but of this we shall speak again in another place. Soon after casting such wheels it is advisable to open the mould, and remove the sand from the central parts, so as to make it cool faster; this precaution saves many castings, not only in this particular case, but in many other instances. Uniformity in cooling is as necessary to success as good moulding. The thinnest parts of castings which cool first, will invariably break; but if a casting cools uniformly, there is no danger of strain in the metal.

Chilled Rollers.—One of the most important cases

MOULDING.

of this kind of moulding and casting in iron moulds, is the casting of chilled rollers. There are some good chilled rollers manufactured in the Western foundries, particularly at Pittsburgh. We will not allude to any particular case, but describe the process of making chilled rollers, generally. The mould for a chilled roller consists of three parts, as shown in figure 32. The lower box of iron or wood is

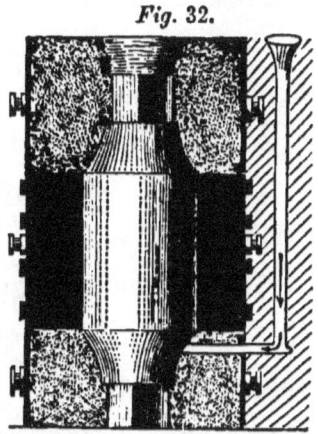

Fig. 32.

filled with "new sand" or a cement, a strong composition of clay and sand, in which a wood pattern is moulded which forms the coupling and the neck of the roller. The middle part of the mould is the chill, a heavy iron cylinder well bored. The upper part of the mould consists again of a box, but is higher than the lower box, so as to make room for the head in which the impurities of the iron, "sullage,"

13 *

are to be gathered. The two boxes with their contents of sand are to be well dried. In many establishments the two ends of the roller are moulded in loam, over the chill, to secure concentricity of roller and coupling; but this can be quite as safely arrived at by fitting the ears and pins of the boxes well to the chill. The chill is the important part in this mould: it ought to be at least three times as heavy as the roller which is to be cast in it, and provided with wrought-iron hoops to prevent its falling to pieces, for it will invariably crack if not made of very strong cast iron. The iron of which a chill is cast is to be strong, fine-grained, and not too gray. Gray iron is too bad a conductor of heat; it is liable to melt with the cast. Iron that makes a good roller will make a good chill. The facing of the mould is blackened like any other mould, but the blackening is to be stronger than in other cases, to resist more the abrasive motion of the fluid metal. The chill is blackened with a thin coating of very fine black-lead, mixed with the purest kind of clay; this coating is to be very thin, or it will scale off before it is of service. The most important point in making chilled rollers is the mode of casting them, and the quality of iron used. Of the latter we shall speak in another place. To cast a roller, whether a chilled roller or any other, from above, would cause a failure, for the roller will be useless. All rollers are to be cast from below. It is not sufficient to conduct the iron in below; there is a particular way in which the best roller may be cast, for almost every kind of iron. The general mode is represented in figure 33, which shows the upper

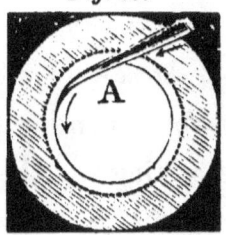

Fig. 33.

side of the lower box. In A is represented the cast-gate and channel, as it is seen from above. The gate is conducted to the lower journal of the roller, and its channel continues to a certain distance around it; it touches the mould in a tangential direction. In casting fluid metal in this gate the metal will assume a rotary motion around the axis of the roller, or, which is the same, the axis of the mould. This motion will carry all the heavy and pure iron towards the periphery or the face of the mould, and the sullage will concentrate in the centre. It is a bad plan to lead the current of hot iron upon the chill, for it would burn a hole into it, and melt chill and roller in that place together. The gate must be in the lower box, in the sand or the loam-mould. The quality of the melted iron modifies in some measure the form of the gate, for stiff or cold iron requires a rapid circular motion, while fluid, thin iron is to have less motion, or it is liable to melt to the chill. The roller is kept in the mould until perfectly cool, but the cooling may be accelerated by digging up the sand around the chill.

Casting Iron to Steel.—One branch of moulding and casting we have to mention before we leave this subject: it is the casting together of iron and steel. At present many anvils, vices, and other articles are made of cast iron, mounted with steel, which are in a fair way of driving all the wrought-iron articles of this kind out of the market. The welding together of steel and cast iron is not difficult, if the steel is not too refractory. This process will not succeed at all with German or shear steel, and hardly so with blistered steel, but it is easily performed with cast steel, by soldering it to cast iron by means of cast-iron filings and borax. Of the manufacture of these cast-iron articles with steel faces we can give but the outlines, having had no opportunity of becoming thoroughly acquainted with this branch. The cast-steel plates to be welded to the faces of anvils, are generally from a half to five-eighths of an inch thick, and as wide as the face itself. These are ground or filed white on one side, and then covered on that side with a coating of calcined borax. The plate, with the borax on it, is heated gently until the borax melts, which covers it with a fusible transparent glaze. The plate in this condition is laid quite hot in the mould, which latter is made of dry and strong sand. The iron is poured in and rises from below; the steel plate being the lowest part of the mould, it will have the hottest iron. The heat to be given to the iron will depend in some measure on the quality of the steel; shear steel requires hotter iron than cast steel. The cast iron used for these purposes, is to be strong and gray, but not too gray, or the union of the iron and steel is not strong. White cast iron will not answer in this case, partly because the casting would be too weak, but chiefly because the cast iron would fly or crack, in hardening the steel. The hardening is done under a considerable heat, with an access of water falling from an elevation of ten feet or more.

Moulds for Lead, Tin, &c. &c.—Besides these moulds of sand, loam, and partly iron, there are moulds which are entirely constructed of metal, either of iron, copper, brass, or bronze. Such moulds are used for casting tin, lead, pewter, Britannia metal, zinc, types, and other articles of economy and ornament. Brass or bronze moulds are generally preferred to iron moulds, because they do not corrode as iron moulds do, and retain a more perfect polish. Such moulds are constructed on the same principle as sand or loam-moulds. If a metal mould is divided into two, three, or more parts, each part is provided with a handle sufficiently long to protect the hands against the heat of the mould. The parts of such a mould must be nicely fitted together, and kept in their position by ears and pins, or in many instances by wedges. The mould is gently heated before any metal is poured into it, to secure the filling of the space in the mould, for many of the most fusible metals and alloys cannot lose much heat from melting to congealing. The moulds must be well polished after each cast, and are then rubbed over with a rag containing oil or tallow, and which spreads a thin film of oil or tallow over the facing of the mould. In many cases a covering or film of pounce-powder—sandarach—beaten up with the white of an egg, is preferred, particularly for alloys. Single metals work better with oil or fat.

Moulds for Copper and Brass, if it is intended to

make sheets of these metals, are for the first metal simply cast-iron boxes, in which the iron is from one and a half to two inches thick. These boxes are formed so as to be taken apart, for the copper will adhere to the iron if it is very hot. These iron moulds are to be very clean, or the cast of copper, which is from two to three inches thick, is apt to have holes, which makes it useless for sheets. Brass may be cast in the same way as copper, but it is more safe to cast brass plates for sheets between two stone-plates. These stones may be of granite, freestone, or any other kind of hard fine-grained quartz stone. They are to be from six to twelve inches thick, and secured against falling to pieces, in case they crack, by iron hoops. The space between the stones for making the thickness, is formed by iron rods. Such a mould is to be in the sweep of a strong crane, and is in the whole a somewhat complicated operation, foreign to our subject.

Stereotyping.—*Plaster of Paris* moulds are used for many articles cast of fusible metal, but particularly for stereotyping plates used in printing books. Fine plaster of paris is first cast over a page of letter composition, and this thin coating strengthened by coarse plaster. This plaster mould is dried at a boiling heat in appropriate stoves, and then dipped in a kettle filled with melted type metal. When the mould is cooled the plaster is broken off, and, according to the skill of the operator, a more or less true copy of the letters which served as a pattern is obtained.

There was a kind of stereotype process formerly practised, which deserves, on account of the principles involved, not to be forgotten. Before the invention of the present mode of casting stereotypes under the influence of pressure in a metallic bath, they were made simply by pressing the pattern,—which might be a wood cut, or a composed form,—upon the liquid metal, just when at the point of congelation. It was a process which required skill and dexterity, but made better casts than the present mode of stereotyping. The fine stereotyped prints made at the end of the last and the first part of this century were stereotyped in this way. The beautiful stereotypes of Firmin Didot in Paris were done in this manner. The metal used for making the mould was lead with a little tin; this was melted and cast in a paper-box as large as the cast was to be. The fluid metal was but one-eighth of an inch thick and resting upon a level table, cooled very uniformly. The moment when the metal was going to crystallize (assume its solid form) was the time to put the wood engraving or form of types down upon it, with a certain force. This process, performed with skill, made better and more correct impressions than the present plaster of paris mould. This first or lead impression served as a mould for the next cast. The next cast was made of type metal, or an alloy still more fusible. This metal was cast like the first, in a low paper box, and the moment when it was going to congeal, the lead mould was with force put down upon it. This latter cast was the true copy of the pattern. The paper boxes were surrounded by a screen of sheet iron, to protect the operator against the flying hot metal. The thin film of oxide, covering the melted metal, was the means of preventing the adherence of one metal to the other. Machines have been in operation to perfect this process, and make it less dependent upon the operator; still, the present mode of casting stereotypes has prevailed over the old method, as it is supposed to be more advantageous. If there is no advantage in stereotyping letter-press in the old way, it is certain that engravings are made more perfect in that manner. The composition of the metal for this art, may be varied from the melting point of lead to the melting of an alloy which requires but the boiling heat of water.

Impressions and Castings.—Before we proceed to the consideration of metals, we will speak of som. interesting operations connected with the fine arts. We allude here only to relief impressions, not to those in ink or colours. The materials in which impressions may be made, are wax, paper, whalebone, horn, glass, sulphur, and many other materials to be mentioned in the course of this chapter. Impressions are made in many materials, and a variety of operations in the useful arts depend upon this manipulation. The operations in the mint, and stamping of medals and utensils, as spoons, forks, and pans, are parts of this branch of art; reliefs in copper, brass, and silver sheets, the pressing of wooden snuff or other boxes, of handles for canes and umbrellas, of leather, cloth, and paper, all belong to a different branch from that we are investigating. Most of this work is performed by stamping-machines and dies, where the relief part of the die is stationary, and the counterpart or intaglio moveable. Some of these operations are closely connected with our art, and for these reasons we will describe a few of them. Impressions of small objects are easily taken: the difficulty in making large impressions increases rapidly with the size of the impression. The use of impressions in this case, is to obtain moulds from patterns which will not bear a cast or mould, as coins, gems, &c.

Wax is one of the best materials to take impressions with; yellow wax is particularly qualified for this purpose. Before using it, it is to be gently warmed and worked between the fingers, after which it is more uniform in composition, less adherent to other matter, and stronger in itself. The only objection to it is, that it is not very durable, and is to be kept with caution to save the sharp impressions of the original. Such impressions in wax are made where the original pattern will not bear heat or water. Their use is to make plaster coats over them, and prepare the plaster cast for patterns to be moulded in sand.

Bread in crumbs, is another material for taking impressions. If this is well worked between the fingers before the impression is taken, it can be dried without cracking, and casts of sulphur, plaster, or other matter may be made in it with success.

Impressions in *sealing-wax* can only be made in cases where the pattern is not liable to injury from the heat of melted sealing-wax. In this operation sealing-wax of the best quality is required; it is to be melted in a thin layer in a metallic capsule over the flame of a lamp, and the pattern, as lapidary or seals, is impressed upon it when near the point of congelation. Impressions in sealing-wax are very useful for taking casts in clay or plaster, and if enclosed in a metal capsule they may be moulded in sand. The melted wax must be free of blisters, and the pattern which gives the impression very clean.

Sulphur, is a material very useful in taking impressions, but it is somewhat difficult to succeed with it. There are two ways in which it can be done: we will mention both. If sulphur is melted to nearly its boiling point, it assumes a pasty appearance. If in this condition it is quickly cast into a large vessel of cold water, it will retain that pasty form. The detached parts may be united under water, without injury to the condition of the sulphur. This putty sulphur will take fine impressions, and regain in a few days its natural hardness. A less difficult operation is the following. In melting sulphur it first assumes a watery appearance, is clear and liquid, but by increased heat becomes brown and tough, and at last it burns with a blue flame. In this state it is cast upon a plate, where, in gradually cooling, it becomes liquid, and after this congeals all at once. When the sulphur is just beginning to harden, the pattern is pressed firmly upon it, and a good sharp impression is thus obtained.

Glass impressions are very durable, but are not so easily made. To copy a coin, cameo, or medal in glass, an iron welded ring about a half or three-fourths of an inch high, a little larger than the pattern, is laid around it. In this iron ring upon the pattern, damp tripoli of Corfu,—other kinds of rotten stone cannot be recommended, because the chemical composition of this tripoli is the chief condition of success—is rammed on just as in sand moulding. The facing is to be the finest part of the tripoli, and worked through a fine silk sieve. When the pattern is removed, this mould is at first gently dried and gradually exposed to a stronger heat, to expel every particle of moisture. Upon the face of this mould a round piece of fusible glass is laid a little larger than the pattern, and the whole exposed to the heat of a cupola or muffle, such as assayers use for refining and assaying. The glass will soften by degrees and fill the mould, the refractory character of the silicious tripoli preventing it from melting together with it. Coloured impressions may be made simply by melting the coloured glass first down into those parts which are to be coloured, and then covering the whole with such glass as we intend the body of the impression to consist of. This latter process, however, requires two moulds, and two opera-

14 *

tions; the first mould makes but one colour of glass, which is to be ground on its reverse, before the second or body plate can be melted to it. The glass used in this art is that of which pastes or artificial gems and precious stones are made.

Clay is an excellent material for taking impressions, but its shrinking, and consequent cracking, make it less useful as a material for taking impressions. It is most extensively employed as a means of raising ornaments upon porcelain. If coloured ornaments are wanted, the white clay is coloured by a fire-proof colour, pressed into a bronze mould, made flush with the mould by a bone spatula. The ground mass is laid over it, to which it will adhere. The contraction incident to clay impressions may be brought to useful account. By repeated moulding and drying a diminution of the original pattern may be obtained, true in all particulars, but somewhat less sharp.

Artificial-wood impressions may be made by mixing saw-dust with a solution of glue 5 parts and isinglass 1 part. The moulds for this mass may be made of metal, wood, sulphur, or even plaster of paris, covered with a film of oil. The mass is pressed into the mould by hand. Impressions of this kind are never sharp, but answer for many purposes instead of wood-carvings. They may be varnished and gilded like wood, but cannot be used in damp places. Saw-dust of willow, maple, gum, and similar kinds of wood, is preferable to that of hard wood, as mahogany, or pine wood. An addition of finely powdered chalk, rotten-stone, or fine sand, improves the sharpness of the impression. Clay does not answer in this composition, on account of its affinity for water.

. Castings of other materials than metals are not extensively in use, but are of importance as means of making patterns.

Plaster of Paris is the most important in this range of materials. It is made by calcining pounded or ground gypsum gently in an oven: a common bake-oven is sufficient for a small quantity, for there is no other ingredient in the composition of the gypsum to be driven off but the water of crystallization. Too much heat deadens the plaster, and too little heat makes it work slow and absorb less water of crystallization. Plaster of paris exposed to atmospheric air loses its quality of hardening with water; gentle heat in an iron kettle and stirring, restores the lost capacity for water. To work successfully in plaster, experience and skill are required, but we will try to give as good practical information as is in our power.

One of the first requisites to success in this work is a thorough acquaintance with the nature of the plaster. If the material is a strange one, it is advisable to calcine it in an iron kettle under repeated stirring to a red heat, or so far as the kettle will admit of, before running the risk of a cast. The quantity of water with which any kind of plaster will assume its greatest hardness, is to be tried by experiments. Some qualities absorb more water than others. The hardest casts are made with the least water, but it requires dexterity to make sharp castings of a stiff pasty plaster. The casts are also harder if warm water is used. To prevent large pores, and blisters in the cast, the solution is to be constantly stirred, and kept in motion until the plaster is hardened in the mould. The best plaster casts are made if a very thin solution is first spread over the face of the mould, and upon this, while wet yet, a stronger cast is made. This will unite strength and beauty in the same cast. Foreign matter ought not to be mixed with plaster: it invariably impairs the strength of the cast. If plaster is to be used for making patterns, one-third of slacked lime may be mixed with it. This keeps the plaster for a long time in a pasty condition, and offers an opportunity to alter the form of it so long as it is in that state. A little lime mixed with pure plaster, makes it more useful for moulds, particularly where metals are to be cast in it. The best mixture for making moulds of plaster for metal, is to mix it with one-third of finely ground pumice-stone, and a little clay. All other admixtures to improve the hardness or strength of plaster are useless. The strongest casts are casts of fresh, well burnt plaster, which was not too thin when cast. A mould of plaster may be made over any pattern which is impervious to water; therefore all patterns which absorb water are to be covered by a varnish which excludes water. In varnishing a pattern the varnish is to be laid on thin, and uniform, not to mutilate the pattern, or fill up fine cavities. As an illustration of this subject, we will give a description of some practical cases. To cast a mould of a coin, or of a wood engraving, the pattern is first brushed over with oil or soap-water, and then laid on a level place upon a board or table. It is now surrounded with an enclosure of varnished pasteboard, tin-plate, or anything light and flexible, which is to be fastened tightly around the pattern. This is to project above the face of the pattern the proposed thickness of the plaster cast—if it is higher there is no harm done. Plaster of paris is now mixed with an excess of water, in a common water pitcher, well stirred, and after remaining a moment at rest, the coarse plaster will settle at the bottom, and the finer portion be suspended in the water. The lighter part of this liquid is gently cast over the pattern, while the latter is constantly and gently struck, so as to settle the particles of plaster in the finest crevices of the pattern, and make air bubbles rise, which often pertinaciously adhere to the pattern. The coarse sediment of the plaster is thrown away, or saved and exposed to another fire before being used again. After five or ten minutes' standing, the fine plaster is settled in the mould, and clear water stands over it. This water is cast off as dry as possible, and some fresh plaster, mixed very stiff, is cast over the first thin facing to strengthen it. The first cast is made very thin merely to cover the pattern, for it will be too weak and porous for any practical purpose, even if cast thicker. The two casts will unite firmly, and form a useful whole, giving a very minute impression and being strong besides. Such a plaster mould is dried, to expel all the water from it, and may then be used to cast fusible metal, wax, or sulphur in. If this mould is to be used for making plaster casts, it is varnished first, which is done by a gum-shellac varnish, or by soaking the mould in wax. The first is the preferable plan. The first coating or facing of plaster may be put on by a fine camel's-hair brush, but this way is not so sure of making perfect impressions as that described. There is a certain time for removing the cast from the pattern; if this is done too soon the cast is too soft, and will break, and if done too late it will adhere to the pattern. For small objects, and strong plaster, from ten to fifteen minutes is sufficient; for larger ones, from fifteen minutes to one hour will be required, before the cast can be separated from the pattern. The patterns are to be covered by a film of oil, as remarked before; this subject requires more attention than at first sight appears necessary. Pure oil is liable to fill the finer parts of the pattern and prevent the access of the plaster; it has, besides, the evil influence upon the cast that it prevents the hardening of it, and if, therefore, the cast is sharp at first, the least rubbing will abrade the facing, at least the finer parts of it. A solution of white hard soap brushed over the pattern is preferable, but if the pattern is not very well smoothed or well varnished, if of wood, the cast is apt to adhere to the pattern. In most cases a mixture of a strong solution of soap.

and a little oil, is found to be the best parting material. Oil generally gives a colouring to the white plaster, white hard soap does not.

The Moulding of Statues in plaster of paris is not an object of general interest, and for this reason is hardly worth the pains of describing and reading an essay on it; but as it affords the best illustration of moulding busts and statues, we will give this subject more attention than we otherwise should do. There are three different ways of moulding a complicated statue. The first is to make the mould and the cast in parts, and screw or cement these parts together. This is an imperfect mode of forming statues, which never makes correct work, for it depends not only on the moulder, but also on the finisher who puts the parts of the statue together, how far the cast may be true to the original pattern. The parts of metal statues are screwed together; if plaster they are cemented together by plaster, and the joints smoothed. Statues of this kind are weak, nor can they be correct, as it is almost impossible to destroy all traces of the joints.

The second manner of forming statues is to cover the original with a thin coating of plaster, one-fourth to one-half of an inch thick, and paint this coat black,

MOULDING.

giving it a very thin film of charcoal-powder, strengthened with glue, and over this coating a thick coat of gypsum, two or three and more inches thick, according to the size of the pattern. This is laid on with the trowel. When this last coat is sufficiently dry to admit working at it, the cope is divided by black chalk into so many parts as are necessary to secure the separation of the cope from the pattern. The moulder of course is to be well acquainted with the pattern, or he could not with any certainty mark the parting-lines on the cope, having no means of ascertaining and tracing the lines on the pattern. To make this operation less difficult, a part of the pattern may be left uncovered, say the back (of a statue); this makes the tracing of the partings more safe. The omitted part is covered in a second operation, where the joining is formed by that line, and those parts of the cope which enclosed the covered space. The partings are effected by cutting down with a chisel or saw through the cope to the black stratum, and breaking the first covering of the pattern. The black paint forms here a uniform stratum interlining the cope; it gives warning to the operator to stop cutting, for the pattern is near. This mode of operating is easy and safe, as it makes a good and correct mould; but the broken edges which form the parting are very soon injured, and show unsightly joints on the casts. For plaster this method is imperfect, because it does not make many good casts. One cast may be made very correctly, but the following casts are not certain. The parts of the mould are held together by winding tape or twine around the mould.

The third plan of making a plaster mould is tedious and slow, but is the safest and most correct, and by good treatment of the mould may admit of sixty and more castings being made in it. The manner of forming such a mould is the following, which, with unimportant modifications, is practised in making moulds for metal casts. The surface of the pattern is marked by a lead pencil with such divisions as will secure the lifting of that part of the mould from off the pattern, as is enclosed by such marks. The operation of making the mould commences on a convenient place, by enclosing one division with fine plastic clay, and giving the borders towards the enclosed space that form which will cause the plaster to have the shape desired for that particular spot. The space enclosed by the clay is then filled by plaster, and when the latter is settled, and so far dried as to admit its removal, the clay enclosure is first removed. This leaves a part of the mould to be made, or the plaster cast standing. This cast may be one, two, or three inches thick, according to circumstances, it being the object to equalize the surface of the mould, so as to have less abrupt reliefs. This first part of the mould is taken off the pattern, and the edges cut smooth by a knife. The taper of the edges is so calculated as to form the joints of an arch, so that when all parts of the mould are laid together without the pattern, no part of it can move or fall off from the others. To secure the relations of the parts of the mould still more perfectly, each part is provided with warts in the joints, fitting into opposite hollows of the next part. These warts are made with the point of a knife, by turning it backward and forward, and are set in the middle of the joints, or in such places as are considered more convenient than the middle. When the first part is dressed, it is again put in its place, and one side of it joined by clay enclosures. If the space now to be covered is square, the plaster will form one side of it, and the three other sides are formed with clay. This second space is again filled by plaster, and it forms part No. 2 of the mould. One side of No. 2 fits to one side of No. 1, and three are to be dressed and provided with hollows for warts. In this way the whole pattern is covered with small parts of the mould, which in many cases require fifty or more cores or parts. The last part of course is cast without any clay to form the enclosure, and is generally without warts to form the starting point in separating the mould. When the pattern is perfectly covered with this mould, the surface of the mould is dressed and cut smooth, to remove all sharp angles and abrupt reliefs. Over this first cope is made a second cope, but the first ought to be in such a condition that the second divides only into two, or at most into three parts. The divisions of the first cope of course fit exactly into the second, and if there is any doubt or danger that one of the parts of the first cope would fall out from the others in turning the mould, that part is to be provided with a wire staple to which a string is fastened. This string passes through the second cope and is secured outside. The second cope may also be provided with warts which fit in corresponding holes in the first cope, if found necessary, which, however, is not often the case. The whole mould, forming a comparatively heavy mass of plaster, is held together as in other cases by means of tape.

Large Plaster Castings are made hollow. This is done by casting first a small quantity of fine plaster in the mould, and in turning, the mould is led into all parts of it, and gives a thin covering to the whole face of the mould. A second cast of coarse plaster follows the first soon after, and this is equally distributed over the mould. A succession of such casts will give any thickness desired. Parts which require extra strength are laid on by hand or the trowel. Statues and busts generally require no castgate, because they are open below and are cast from that side.

Patterns and moulds in which plaster casts are to be made, are coated with a film of oil or soap; but valuable pieces of art, as marble statues or busts, do not admit of oil or soap without injury, and these means cannot be employed. In such cases the pattern is covered by tea-chest-tin or tin foil, but so as not to show the joints of the foil. The tin-foil is pressed on by a cloth-brush in such a manner as to secure the perfectly close covering in the undulations

The face of a living or dead person may be copied in plaster by making a plaster cast over the face. The limits of the mask are marked by laying a wet cloth around the face. The hair and eyebrows are covered by pasting some tin-foil over them. Living persons are to have two paper or tin-plate pipes in the nose, to admit of breathing while the plaster is put on the face. Such

15 *

masks are generally used as patterns for making busts of those persons from whom they are taken. The hair, ears, and the back part of the head, are to be supplied by the artist.

Sulphur is, next to plaster of paris, the most valuable material for sharp castings; but its application is limited to very small castings, on account of its brittleness. It can be cast over metals and many other materials without oil, and gives for these reasons very sharp impressions. Sulphur may be cast over a coin by surrounding the coin with a ring of paper; the melted sulphur will not kindle the paper if it has the proper heat. In melting sulphur for casting, it is not to be overheated; at first heat it melts to a transparent clear fluid, and that is the time to cast it. More heat transforms it into a pasty mass, which cannot be used. The kindling of the sulphur should be prevented, by all means, for it will impart a dirty gray colour to the sulphur. Sulphur may be mixed with foreign matter to augment its strength. One part of plaster of paris, and two of sulphur, improve the tenacity of sulphur without diminishing its capacity for fine impression Next to the above, fine Spanish brown, fine chalk, or clay in powder, may be mixed with it. Three parts of sulphur, and one of silver, is a good composition for sharp and durable impressions.

Wax in its pure state, as well as mixed with other matter, is a useful material for castings, but it shrinks considerably. It requires skill not to cast it too warm, or too cold. In the first case its castings will be defaced, in the latter they will not take sharp impressions. Wax may be mixed and successfully used with plumbago, cinnabar, white-lead, plaster of paris, and other substances. The mould wherein wax is to be cast, is to be very cold or wet, if the material admits of the absorption of moisture. When the face of the mould is covered by a thin coating of wax, the surplus fluid wax may be cast back into the ladle. A thin cast will not shrink so much as a thick cast.

Sealing-wax, isinglass, and glue, are also materials for making casts of, and are frequently used for small articles. There is one composition to which we have to allude more particularly; it is a composition used in making elastic moulds, for casting in plaster of paris—eight parts of glue, four parts of molasses, mixed and boiled together, and to this gradually added one part of varnish or boiled linseed-oil. This mass is cast hot over a pattern, and when cooled may be easily removed. It forms a gelatinous mass, and makes an excellent mould for plaster casts, having the great advantage of admitting of under-carving the pattern. Such a mould will not make more than six or eight sharp casts; but as the making of the mould is no object, it is the cheapest and quickest way of forming a mould for casting plaster in.

Alum cautiously melted, so as not to expel its water of crystallization, will assume a very fluid appearance, and may be cast in small moulds with success. Thirty parts of alum and one of saltpetre is still better; it makes opaque castings of a beautiful white. Five parts of alum and one of common salt melted together, makes transparent sharp castings. Melted saltpetre by itself, may be cast in hot metallic moulds, and makes castings of a fine alabaster appearance.

Moulding Natural Objects.—A mould over an object of nature, as over a small animal, a flower, or leaves, may be made in the following way. The dead animal, say a fly, or a bug of any kind, is put with its feet upon a ring of wax, so as to place the feet and everything else in such a position as we want it in the cast. This wax ring will form the channel or gate for the fluid metal. The object—animal or leaf—is painted with a very thin solution of gum-shellac in alcohol; and, after being dried, is placed in a small pasteboard box, and so fixed by means of small wires as to secure it in a permanent position. These wires, after being withdrawn, form air channels through the mould. A small tapered pin of wood is fastened in some convenient place for making a cast-gate. A mixture of three parts of fine plaster of paris, and one part of fine brick-dust, formed by an adequate amount of water, to which a little alum and an equal portion of sal-ammonia is added, into a thin pap, is now gently cast over the pattern, under continued shaking of the mould, or if that cannot be done because the pattern is too delicate, the pattern may be first covered by means of a fine camel's-hair brush, with a thin coating of the above mixture, and then the remainder cast over it. When this cast is hardened, the pasteboard enclosure is removed, and the cast gently but very strongly dried. After all the water is expelled, the mould is brought slowly and gradually to a cherry-red heat, to expel and burn all the animal and vegetable matter. A mould of pure plaster would not resist such a heat without falling to pieces, but an addition of brick-dust and alum gives it that resistance to heat which is needed. The sal-ammonia is added to facilitate the destruction of the natural pattern, the animal or plant. The cooling of the burnt mould is to be performed equally as slowly as the burning itself, to prevent its breaking. In the cooled mould some mercury is cast and gently shaken. By gradually adding more quicksilver, the remains of the pattern which may be left in the mould will float on the mercury, and may be brought out. By repeating the latter operation, all impurities may be effectually removed. Before casting any metal in this mould it ought to be heated to a certain degree, which degree will depend in some measure on the mass of the pattern, and the metal to be cast in it. Very thin fine patterns, and metals which congeal quickly, require a hotter mould than the reverse qualities. Silver is the best qualified for such casts: after this, type metal, tin-solder, and fusible alloys. A cast made in this way may be prepared to form a pattern for the current business of the foundry. If the mould has been hot and the metal also, the casts are generally so perfect as to show the finest nerves of the pattern. Larger objects may be moulded quite as successfully as small ones, but it requires more experience to succeed as well.

CHAPTER II.

FOUNDING.

MELTING OF METALS.

Iron.—It is impossible to qualify the various kinds of pig-iron brought into the market, by local terms and marks. It would, after all, not be of any use, because the produce of one and the same furnace may change in one week's time from No. 1 iron to No. 2 or even No. 3, which certainly makes a great difference in its application in the foundry. There are, however, distinctions in the quality of iron caused by the ore, or by the fuel which has been used in its manufacture, as charcoal or anthracite; as well as by manipulation. We will allude to these local and practical differences when pointing out the specific qualities of metal for certain purposes, and confine our demonstrations at present to general remarks. Taking no notice of the difference between charcoal, anthracite, and coke or stone-coal iron, we have three distinct qualities, known as No. 1, No. 2, and No. 3 iron.

No. 1, *or Dark Gray Pig-Iron,* is the foundry iron.

This pig-iron is, if anthracite and charcoal, mostly of a coarse-grained, apparently crystalline fracture. There are, however, no crystals; the form of the fracture is an aggregation of leaves. Iron, and the black graphite with which it is intermixed, appear to assume the same crystal form; they are so closely united that no distinction can be made of the difference in the form of the crystals, if there is any. Coke-pig, No. 1 stone-coal iron, and hot-blast iron, are generally finer in the grain than the above-mentioned qualities. Pennsylvania anthracite pig No. 1, and Pittsburgh or Hanging Rock No. 1, are generally very coarse and black in the grain fracture. Charcoal iron No. 1 from the Eastern States, Maryland, Allegheny river, and Ohio river, Tennessee and Kentucky, is generally hot-blast, and finer in the grain than the above. Scottish pig, is of a fine-grained fracture.

The pig-iron of this class is soft, and often tender: most of our own manufactured iron is strong. It melts very fluid, and cools very slowly, which qualifies it particularly for castings. This iron, if very gray, may be remelted once or twice, but the fine-grained kinds, and those which contain less carbon, or are exposed to too much fresh air in melting, turn into the following, or

No. 2 Iron.—This contains less carbon than the above, is more gray in appearance, and of a finer grain. If approaching near to No. 1, it is the best foundry iron, for it is stronger than No. 1. If this iron assumes a more gray colour, it is not qualified for small castings, but is very excellent for large castings in dry moulds. It melts fluid, fills the mould well, makes less sullage than No. 1, and does not burn the mould so much as the above. It is tenacious, may be filed, turned, planed, and polished; it is close, and more certain to be free from impurities than No. 1.

No. 3, is white pig-iron. By remelting No. 1 and No. 2 under the influence of a liberal access of air, they will be converted into No. 3. This iron is white, and most of it of a bright crystalline fracture. It is of no use in the foundry.

The quality of foundry-pig in our Atlantic cities, also in Pittsburgh, Cincinnati, and other cities along the western rivers, is no doubt of such perfection that there is no difficulty in making any quality and kind of castings in any of these places. There is hardly a limit to the variety of good foundry-pig in these markets. Some general remarks on the characteristics of pig-iron for foundry purposes will however be in place.

Dark Gray pig-iron, with large leaves of plumbago

is qualified for small castings, as hollow-ware and small machinery, but would not answer so well for heavy castings, which require strength. There are, however, exceptions to this rule. The pig-iron most useful for the very finest kind of castings, is to be fine-grained. Coarse-grained pig will not fill a fine mould, at least will give but dull impressions. If pig-iron contains a little phosphorus, it may be fine-grained and still be an excellent foundry iron, particularly for hollow-ware and stoves. Hollow-ware made of gray iron which contains much carbon or plumbago, is liable to cooking black; this evil is not so apparent where pig-iron of lighter colour, containing a little phosphorus, is used. Black iron is not qualified for large or heavy castings, as it is generally too spongy.

Hot-blast and cold-blast iron are simultaneously brought into the market, and the former is frequently sold for the latter. For foundry-pig it makes but little difference whether made with hot or cold-blast, and we may say, generally speaking, that hot-blast iron is preferable to cold-blast, because the grain is finer, the iron more uniform, and it runs more fluid than the latter. In anthracite and stone-coal pig there is but one kind, and that is hot-blast. A difference is often found in charcoal-pig, but then it is generally marked cold or hot-blast, when made at an establishment of reputation. To distinguish cold-blast from hot-blast iron, is almost impossible. The only permanent difference is a finer grain in hot than in cold-blast, provided the amount of carbon in both kinds of iron is the same, and the iron is made from the same kind of ore. This mark of distinction is, however, very doubtful, and may lead to errors. A more certain criterion is the colour and lustre of the pig, in a fresh fracture. Provided all other things—as ore, coal, manufacture—are equal, the fracture of hot-blast iron is duller than that of cold-blast; the latter shows more life than the first, and a freshness of colcur, which is not so clearly expressed in hot-blast iron. Hot-blast iron is frequently found to be of a fine grain, interspersed with clusters of coarse grains, the fine parts of a dull appearance. These distinctions of colour are a safer criterion than the size of the grain, but both together may afford some means of distinguishing between the two. It would be of little value to know whether a specimen of iron was smelted by hot or by cold-blast; but as the cold-blast iron contains less carbon and impurities, if of the same colour as hot-blast, and as a mixture of cold-blast and hot-blast iron makes the strongest castings, it is desirable to have the two qualities separated.

The mixing of different kinds of iron is an object of considerable interest, and all foundries ought to make their own experiments to ascertain the strength of the material they are working. In making ornamental casts, strength is of secondary consideration, but in machinery, and beams for architecture, it is of the first importance. In foundries where machinery is cast, or water pipes or beams for bridges or architecture, there should be means of testing the strength of their cast-iron. The safest and best way of doing this, is to have a standard pattern, say a prism of two feet long, one inch thick, and two inches wide. This pattern is to be moulded in a particular flask, with uniformly dry sand, and cast inclined at a particular degree. The mixture of iron is made in a crucible melted in an air furnace. This trial or proof-bar is fastened with one end in a vice, and at the other end a platform is suspended, upon which so much weight is piled as to break the bar. In the mean time the deviation from the straight line, or from its original position, is measured. In this way the relative strength as well as the degree of elasticity may be measured, and the relations of the strength of one mixture of iron to the other, decided on with great certainty. This is not to be considered a scientific experiment—it is a mere matter of local, practical interest. Under all conditions, a mixture of iron melted together is stronger than the average strength of the whole, each measured by itself. Hot-blast iron has the advantage of being of a more uniform texture than cold-blast iron, and being more firmly united with carbon. A mixture of hot-blast iron may therefore be made which supersedes any cold-blast iron, in respect to strength, provided hot and cold-blast are made of the same materials, and in the same manufactory. The kinds of pig-iron which are to be mixed together to form the strongest compound, are difficult to decide upon here. It depends very much upon the experience of the founder, and also on circumstances which are beyond his control. Few of our blast-furnaces have yet settled upon a definite quality and mixture of ore, shape of the furnace, and other matters which influence the quality of the iron smelted. So long as such matters are not definitely settled, no brand of pig-iron can be depended upon for its quality. In purchasing, the buyer is to depend upon his own experience and chance. If pig-iron is too gray, or too spongy, it may be improved by adding No. 3

iron, or in most cases scraps of old castings are preferable. Very black-gray iron will bear an addition of 30 per cent. of No. 3 pig or scrap. Iron which contains too little carbon is successfully improved by adding No. 1 until the wished-for strength and texture are obtained. In all cases iron from different furnaces ought to be mixed together, and if there is any possibility of obtaining iron from different localities and different ores, it is to be preferred. An anthracite pig of the Schuylkill region is stronger if some Scottish pig is added to it; charcoal iron from the State of New York, or from Baltimore, is still better for that purpose. The superior qualities of Ohio iron may be made still stronger by mixing it with some kinds of Allegheny or Tennessee iron. In all cases, however, it is better to mix No. 1 of one kind with No. 2 or No. 3, or scraps of another kind. And if possible, mix cold-blast with hot-blast iron. The strength of iron depends a great deal upon the mode of melting it, but we shall speak of this hereafter.

Besides the consideration of strength, economy in many instances decides the qualities of iron to be worked in a foundry. True economy, however, is that which secures the best castings, and gives most security in avoiding scraps. A mixture which makes a close and compact soft gray iron, is the best in all these instances.

An important influence in mixing iron is due to the kind of casting, its size, and its purposes. Iron of which beams and rolls for iron mills are cast would make poor hollow-ware or ornaments, and iron which makes sharp impressions on small articles, is generally not qualified for heavy articles. Heavy machinery is best made of No. 2 anthracite iron, or a mixture of No. 1 anthracite, and No. 3 charcoal. The variety of anthracite iron is not indifferent in this question, for there is some very weak, also some very superior iron. Hanging Rock pig of good quality is no doubt the strongest cast iron in the world, and it would be an advantage to western enterprise if scientific experiments were made to decide the value in numbers of its superiority over other pig-iron. Small castings and ornamental castings require a fusible iron which coagulates soon and is not too gray, so as to assume sharp impressions. Iron containing a little phosphorus, being a little cold-short, is preferable for these purposes; that smelted of bog-ores is the proper kind for small castings. Railings and ornaments which require strength to resist sudden jerks, are to be cast of a fine-grained, pure iron. free from phosphorus or any such admixture. Chilled rollers or chilled wheels require a very strong No. 2 iron, but it is preferable to make No. 2 of No. 1 and scraps or No. 3 charcoal. In hard rollers a little phosphorus does no harm, but in wheels any pig-iron made of bog-ore is to be rejected.

The kind of mould in which iron is cast has a decided influence upon the strength of the cast. Machine frames, beams, rollers, and all castings which require strength, are to be cast in dry sand or loam, for green sand will cool the cast too rapidly, and cause it to chill, or become hard and brittle. Castings which ought to have a good smooth surface, to be perfect, require a green-sand mould. A mould well dusted by blackening will make smooth and good-looking castings. Thin castings, that is, castings which soon cool, are always more smooth than those where heavy masses of metal are confined to a small space. Castings which require strength are to be cast upright, or at least inclined, having the cast-gate to enter from below, and a flow-gate at the highest part of the mould.

MELTING OF CAST IRON.

Iron in the Blast-Furnace.—Iron is in some few instances used directly from the blast-furnace to make castings of. It is done in those places where fusible ores, as bog-ores and hematites, are smelted by charcoal in small blast-furnaces. There are but few establishments where this is practised; some are along the Atlantic sea-coast, a few in the interior of the Eastern States, and but very few in the Western States. The whole business done in this way does not amount to much. There is really no advantage in casting directly from the blast-furnace, for the iron is never of such uniform quality as to secure good castings. It is on the whole disadvantageous, and more expensive than remelting the cast iron, and giving it a proper quality by mixing it with other kinds of iron. There are, however, instances where casting from the blast-furnace is not only excusable but necessary. Where bog-ores are smelted which make cold-short iron, it is advisable to transform the iron directly from the blast-furnace into castings. Iron, cold-short of phosphorus, is generally not used in forges, and it has too little carbon to admit of remelting. There is hardly any other way left but to make castings of such iron. It is not qualified, however, for machine frames, or castings which ought to be strong. The only and best purpose it is adapted to, is for casting hollow-ware and stoves; it will form fine and sharp castings, and cooking pots made of such cold-short iron cannot be surpassed in quality. It makes enamel superfluous. The usual way of casting from the blast-furnace is to prepare a stopper of slag, just fitting in below the timp of the furnace. This stopper will separate the interior slag and that in the forehearth of the furnace, provided the stopper reaches down into the liquid iron, the blast at the furnace of course being stopped. The surface of the iron in the forehearth, after being cleared of its slag, is clear and will keep so, provided the stopper is thick enough and remains in its place. The iron is dipped, with dippers or ladles of cast or wrought iron, as far as this can be accomplished; after this the stopper is removed, the cinder from the back of the hearth drawn forward, and the furnace put into blast again. A more perfect way of taking iron from the blast-furnace is to make a dip-pool in one of the tuyere arches, provided for that purpose, and where there is no blast-pipe. If the back arch, opposite the work arch, is chosen, the moulding and casting may be carried on very conveniently, without coming in contact with the smelter and his operations. A hole like a tap-hole is here pierced through the back stone, or one of the flanks of the hearth, down at the bottom or near the bottom, and around this hole a round basin is walled up in fire-brick, and well secured in its place by iron binders. This basin need not be larger than to admit a ladle. The hole which puts this basin in connexion with the interior of the furnace-hearth is to be of such a height over the bottom of the hearth as to leave a cover of fluid iron always on it. This pool is filled with some burning charcoal to keep it warm, and as the iron rises in the hearth it will rise in the pool, from which the moulders may dip and take it at any time they choose. When the pool is once thoroughly hot, it requires no charcoal to keep it so.

In figure 34 are represented two ladles. The one is made of cast iron, the other of wrought iron. The latter is preferable for dipping, because there is less danger of its being burned. These ladles are covered with a thin coating of loam, indicated by the dotted lines. A, the cast-iron ladle, receives a

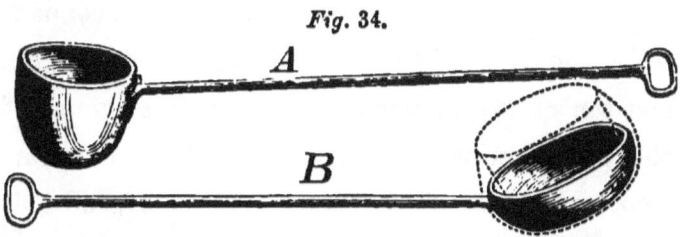

Fig. 34.

strong washing of loam; B, the wrought-iron one, forms merely the bottom to a clay ladle. The well worked clay is set upon the edge of the ladle and forms a dipper as large as the moulder may choose it to have. The clay is put on every day, or every cast, anew, and it is to be well burned before it is dipped into the iron, or dangerous explosions may be the consequence of such neglect.

Melting Iron in Crucibles.—This mode of melting is not now practised, but it was formerly in use, and is still so for some particular purposes. All the fine iron castings, as trinkets and similar objects, are cast from crucibles. The iron melted in a crucible is very quiet, and generally not so hot as to burn the sand; it makes smoother and more solid castings than iron melted in a different way. Compositions of iron may be made and melted in a crucible, which would not retain their quality in any other mode of melting. The melting in crucibles is expensive, because of the cost of crucibles, coal, and labour; but there are instances where these are secondary considerations. A good black-lead crucible ought to last ten or twelve heats of fifty pounds each, and as the plumbago is found in large masses, is cheap, and coal is no object, it may be found a profitable way of making small castings for carpenters and knife-manufacturers. The air furnace for melting iron in crucibles is the same as that used for melting brass, bronze, and similar metals; it is represented in figure 35. This figure explains itself; the furnace is put below ground to a chimney whose lower interior part is built of fire-brick, as well as the interior of the furnace. The furnace is covered by

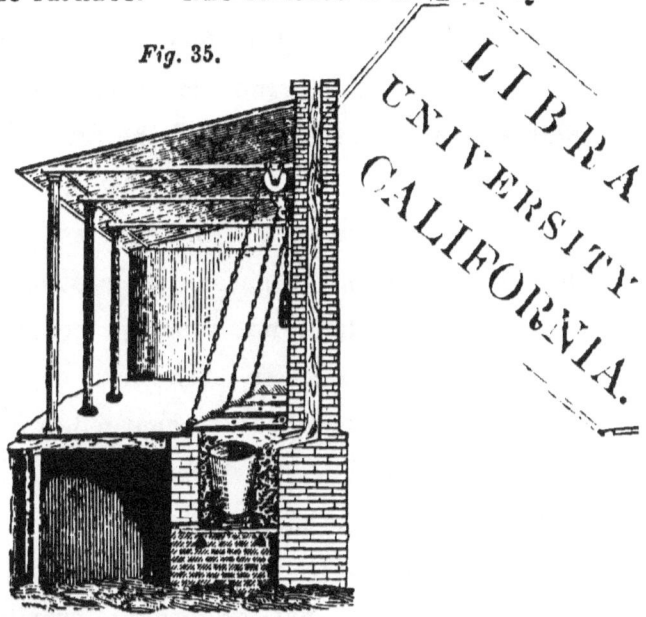

Fig. 35.

a cast-iron plate, a kind of trap-door, which is balanced by a weight and an iron chain passing over a roller; or in any other convenient way. The grate bars are simply square inch-rods of wrought or cast iron, and may be pulled out one after the other, to drop coal and cinders at once, or to clean the furnace. The crucible is set upon a piece of firebrick which rests upon the grate. The bottom of a broken crucible inverted, is preferable to brick as a
17

sole-piece. The crucible is to be raised from three to six inches above the grate, according to the fuel employed. Charcoal requires the highest elevation, coke less, and anthracite the least. The best form for the furnace pit is a square: the four corners resulting from this arrangement are very useful to charge fresh fuel in, which, if the furnace is round, requires more room than can be advantageously given. The crucibles are to be perfectly dry before they are put in the furnace; the least moisture will destroy a crucible if not removed before exposing it to the heat of a furnace. The iron, or other metal, is to be heated before it is charged, and the fuel must be dry and warm, before being laid around the crucible. The mode of operation is simply as follows. The grate is put in the furnace, and upon it the brick-bat or broken crucible, which is to form the pedestal—sole-piece—of the crucible. The fire is then kindled and made to burn briskly, while the crucible and metal are heated on the door-plate. When the interior of the furnace is red hot, and the fuel burnt as low down as the sole-piece in the centre, the empty crucible is put upon it, and then the metal in pieces gradually charged, until the crucible is filled. When the metal is partially melted, there will be room for more, which is piled upon the other, and the whole covered with a few scraps of glass, which, when melted, will form a film on the surface of the iron to protect it against the access of air. A moveable cover of crucible clay will serve the same purpose as glass, but it is more troublesome than the latter. In fifteen minutes the first portions of iron are melted, and the addition may be charged. In three-quarters of an hour the whole of the iron is melted, if properly attended to, and is ready for casting. The fuel is always kept as high as the crucible, and from the first somewhat higher, but the last fuel is given when the metal is not entirely melted, so as not to cool the fire after that by fresh fuel. The fuel is burned down at last so far as to free the crucible of it to a certain depth, and to admit the access of the tongs for removing the crucible. The tongs are made of strong bars of iron, three-quarters or seven eighths square, and and from four to six feet long, one end provided with prongs bent in such a manner as to form a basket to catch the crucible as low down as possible. These tongs are suspended in a chain and a crane, or, the chain very long and fastened to the ceiling of the building. The first operation is to move the crucible from the fire and at the same time put it into a pot-handle for casting. This handle is the same as those on iron pots, to be described hereafter. It is to be heated previously, to prevent injury by cold to the crucible. Two men carry the crucible to the mould and cast, and return the crucible directly to the furnace, into which it is set without delay. Gradual charges of metal are now given, and the melting goes on as before. In case no more metal is to be melted, the crucible is put inversely in the fire to let it cool slowly. In all instances a hot crucible is to be put inversely in case it is set down anywhere; the heated bottom of a crucible never is to come in contact with anything colder than itself. Four or more furnaces may be put at one stack, and as many may be put in a row as is considered necessary. Charcoal may be used in these furnaces, coke is better, but the best fuel is anthracite coal. The danger from the latter is its being too severe upon the crucibles, on account of the great heat it evolves.

Melting in Reverberatory Furnaces.—The best melting furnaces on the large scale are the reverberatories. They are in use in some foundries where the proprietors are desirous of making good castings, but are in a great measure replaced by cupola furnaces. The reverberatory is next to the crucible in making good foundry metal: it gives uniformity

FOUNDING. 197

to the various qualities of pig-iron charged, and the melted iron is quite free from air-holes, and flows like lead into the mould. All founders and engineers agree that castings made from the reverberatory are stronger than those from the cupola, if made of the same iron. In figure 36 a reverberatory furnace is repre-

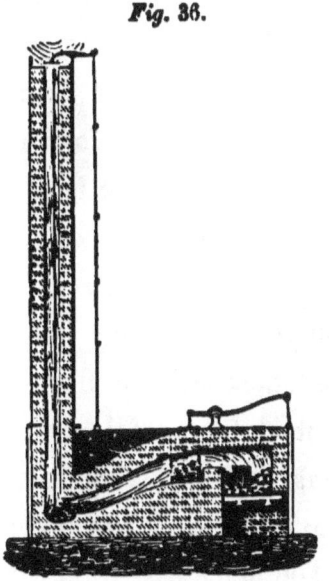

Fig. 36.

sented in section. The whole interior is constructed of fire-bricks, and cemented by fire-proof clay. The enclosure is generally made of cast-iron plates, but we also find furnaces which are enclosed in common bricks, bound together by iron cross ties or binders. The stack is generally 40 and more feet high, even as high as 80 feet; but there is no need of that, as 40

1 *

feet makes sufficient draft. The grate is $3\frac{1}{2}$ feet long and from 5 to 6 feet wide, or as wide as the interior of the furnace. The hearth is from 5 to 8 feet long and equally as wide; it slopes gradually towards the chimney, and forms a basin for the accumulation of the melted metal. The fire-bridge, which separates the fireplace from the hearth, is from 10 to 15 inches high, according to the capacity of the furnace. One side of the furnace is provided with a large iron sliding-door for charging iron and repairing the hearth; this door is at the highest part of the hearth, near the fire-bridge. In the lowest part of the hearth, in the centre of the basin, is the tap-hole. This may be at one side of the furnace, or behind the stack at the flue. A damper on the top of the stack is a useful fixture to regulate the draft. A furnace of this kind is to be very thick in the walls, so as to be as bad a conductor of heat as possible. Too much attention cannot be paid to close joints in the brickwork; open crevices which admit air are to be carefully stopped up, or the iron is liable to a loss of carbon, and will make, in consequence, hard and brittle castings. There are various forms of reverberatory furnaces in use, but the most general is that represented above. There are furnaces with double arches; that is, where iron is

FOUNDING. 199

melted at the fire and at the flue-bridge, and the melted metal concentrates in the centre of the hearth where the arch is drawn down. There are also furnaces where the cold pig is charged in the centre of the basin, which is the centre of the hearth; but none of all these various forms is superior to the above. The pig-iron is here charged behind the fire-bridge, and, as it melts, flows down into the basin. The impure matter adhering to the pig-iron, and which does not melt, as sand and coal, will remain behind the bridge, and may be removed at any time after the heat. In this way, the melted iron is not in contact with any impurities which can injure it. The heat of the furnace is generally greatest near the flue, and the melted metal is in this case exposed to the strongest heat of the furnace. The manipulation at this furnace is very simple. When a cast is to be made at a certain time, the furnace is heated some five or six hours before, and a brisk fire kept all the time; for it will take from three to four hours before the furnace is sufficiently hot to charge iron. The furnace is to be white hot before pig-iron is charged. The large door is then opened and the pig-iron charged, one ton or more at once; in fact, as much iron as is required to make the cast desired; for it is not considered advantageous to charge cold iron while a part is already melted. All the iron contained in a liquid form in the basin, is to be tapped before any fresh pig can be charged. When all the iron contained in the furnace is melted, the tap-hole is opened with a sharp crowbar, and the liquid iron either let into pots or directly into the mould. The tap-hole is stopped with damp sand, or a mixture of loam and coal-dust. When the furnace is charged with iron, all the crevices and joints at the door and in the brick-work are to be cautiously stopped with moist loam, to prevent the access of any air upon the hearth. The firegrate is also to be well attended to, and kept well filled with coal, but not too high, so as to impair the draft of air through the fuel. The grate should be kept free from clinkers, and the formation of holes where the air could pass through unburnt, is to be prevented.

The reverberatory furnace is not only used for melting iron, but is also employed for the melting of large quantities of brass, bronze, tin, lead, and other alloys and metals. Large bells, statues, machine-frames, and similar objects, are cast from the reverberatory furnace. All metals, except very gray, fusible iron, which may be cast from a pot, are to be run in dry sand-ditches, directly from the furnace into the mould. The best fuel for the reverberatory is bituminous coal. Hard coal or coke may be used, but is not so well adapted as the first. The disqualification of the latter arises partly from their incombustible nature, but chiefly on account of the mass of fine ashes which is carried over from the fireplace to the hearth, covering the melted iron and preventing its absorption of heat. This evil is more apparent in the use of anthracite than of coke. Wood, particularly green wood, is not at all qualified for use in the reverberatory; if no mineral coal can be obtained, *charcoal is to be substituted for it.* For the general character and quality of castings, it is to be regretted that the reverberatory furnace for the melting of iron is fast disappearing. Machine-frames of large size, rollers for iron mills, and even chilled rolls, are cast from the cupola. Machine, engine, and iron manufacturers, bridge builders, and architects, ought to insist on having their castings done from iron melted in the reverberatory furnace. Casts from the blast-furnace directly, are the very weakest, and, next to it, ranges the iron of the cupola. The reverberatory and the crucible make the strongest, closest, and safest castings.

The Cupola, has the advantage of melting iron cheaper than any other furnace. Besides this, it is a very convenient apparatus, because a small amount of iron, say fifty pounds, or as large a quantity as five or six tons, may be melted in a short time, with comparatively a small amount of fuel, and in furnaces showing but little difference in size as well as form. In casting small objects, as hollow-ware, agricultural implements, architectural ornaments, and similar forms, and, in fact, in all cases where the strength of the metal is a secondary consideration, there is no question but the cupola is the best form of melting-furnace. There is a great variety in the form of cupolas, but only in minor points; all cupolas generally agree with the form represented in Fig. 37. In A, a section of the cupola-furnace

Fig. 37.

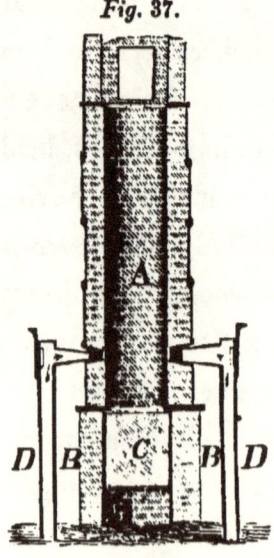

is shown, with another section to represent the sloping bottom. It consists of a cylindrical cloak or enclosure of boiler-plate or cast-iron, of from three to six feet in diameter. This rests upon two brick walls, B B, which are overlaid by a square iron plate, having a round orifice as large as the interior of the furnace. This orifice is closed when the furnace is in operation, by an iron door, C, shut and held close by means of an iron bar propped against it. When the furnace is going out of blast, and is to be emptied of its contents, this door is let down, and with it the slag and hot coal of the interior will drop. The inside of the furnace is lined with fire-brick, or it may be lined with a mixture of fire-clay and river-sand, firmly rammed in and gently dried. A good lining for a cupola may be made of turnpike-mud, where the road is macadamised with flint or hard sand-stone; but, where iron or lime is contained to some extent in such mud, it should be rejected. Some cupolas are but four feet in height, while others are made from eight to nine feet high. We consider five feet as too great a height; there is no other advantage in it than having a larger body of fuel at once on fire, which may be effected to more advantage by a greater diameter. Low furnaces, even as low as three feet, use less fuel than the higher ones. The width of cupolas is quite as variable as the height; there are furnaces of eighteen inches in diameter, and some are four feet. With charcoal, eighteen inches wide and one tuyere will make hot iron, but coke requires at least twenty-four inches and two tuyeres, and anthracite thirty inches or more to produce the same result. A cupola is generally overbuilt by a spacious chimney, to lead the hot gases over the roof of the building; but a sheet-iron pipe will serve quite as well as a brick chimney. The lining of a cupola should be at least nine inches thick, and may be still thicker, if made of fire-brick. These bricks are to be laid in fire-clay mortar, a mixture of refractory sand, and as much fire-clay as is needed to hold the sand together. The tuyeres are generally from ten to fifteen inches above the iron bottom of the furnace, and are simply round orifices, of the size of the nozzle, cut through the in-wall. For small furnaces, but one tuyere is used at the back of the furnace; for larger furnaces, at least two tuyeres are needed; and for still larger, and particularly hard-coal furnaces, we frequently find six or eight tuyeres, cut in the same horizontal plane, in one furnace. If the diameter of the furnace is large, the tuyeres are multiplied, in order to generate a uniform heat at all points in the furnace. Where a large quantity of iron is to be melted at once, tuyeres are cut one above the other; if the melted iron is raised to the height of the lowest tuyeres, these are stopped with fire-clay, and the next above opened, and if the iron is raised to the second, it is also stopped up, and the next higher put in operation. This process is continued until all the iron required for the cast is in the furnace. The vertical distance between the tuyeres is generally six inches. The nozzles of the tuyeres are simply sheet-iron conical pipes, of from three to five inches in width at the narrowest part. The conducting-pipe from the fan to the furnace ought to be at least twice the diameter of the nozzle, or four times as large as the area of all the nozzles. Where more than two tuyeres are used in one furnace, we frequently see a square cast-iron pipe surrounding the furnace; in this pipe are as many orifices, directed towards the centre of the furnace, as there are tuyeres; the nozzles are attached to these orifices.

The operation in a cupola is simple. If iron is to be melted, the first thing to be done, is to lock the iron door at the bottom, then fill in a bottom of sand: moulding-sand is generally used in cases where but a small quantity of iron is to be melted. If a large quantity of melted metal is required, a more refractory sand is used. The fire is kindled by laying a few chips of wood on the bottom, and placing upon them some coke, stone-coal, charcoal, or anthracite. The fire is kindled through the tap-hole, which is at least six or eight inches wide. The tap-hole is left open to admit fresh air for promoting the combustion. The tuyeres are also left open. The furnace is now filled to its mouth with fuel, which is kept at a brisk combustion. It generally requires two or three hours to heat or prepare the furnace for blast, which is not put on until the flame appears on the top of the fuel. When the furnace is thoroughly heated, the nozzles are laid in and the blast-machine is put in operation. Previous to this, however, the large tap-hole is stopped up with moulding-sand, or with a more fire-proof sand mixed with clay, leaving a small orifice at the bottom, which forms the tap-hole for the iron. This tap-hole is $1\frac{1}{2}$ or 2 inches wide, and is formed by placing a tapered round iron bar in the place where the hole is to be, ramming the sand tightly around it, and removing it as soon as the hole is filled up. The blast, when put on, will drive a flame through the small tap-hole as well as out of the top of the furnace. The small tap-hole is kept open to dry the fresh loam or sand more perfectly, and also to glaze the tap-hole so as to resist the abrading friction of the tapping-bar. The flame, also, helps to glaze the lining of the furnace, which is more or less injured after every smelting, and requires mending with fresh fire-clay. When the furnace is to hold a large quantity of metal, the large tap-hole is covered by an iron plate, which is fastened by wedges to the iron enclosure, leaving only the small tap-hole free. The iron is charged as soon as the lower parts of the furnace show a white heat, which is best known by the colour of the flame that issues from the tap-hole, it being at first a light blue, but, with increasing heat, assumes a whitish colour, and apparently a higher heat. In about ten minutes after charging the iron the melted metal appears at the tap-hole, which is now closed by a stopper made of loam, which is worked in the hand until it assumes a certain degree of tenacity; a round ball of it is then fastened on the end of a stick of wood, provided with a disc of iron, which, being previously wet, is then pressed into the tap-hole. A charge of iron never consists of less than two hundred pounds, and, in most cases, of four or five hundred pounds. Pig-iron is broken into pieces of from ten to fifteen inches in length before it is charged. From ten to twelve pounds of fuel are consumed and charged with every hundred pounds of iron in good furnaces. Small furnaces, and those which are driven slowly, use more fuel, and the amount often rises to twenty pounds of fuel to one hundred pounds of iron. Along with the charges of coal and iron, a little limestone, broken into two-inch pieces, or oyster shells, is charged, to about two, or three, and often five per cent. to the weight of the iron. Too much limestone, as well as too little, causes the iron to become white, lose some of its carbon, and in most cases, its strength and softness. The furnace should be kept full while in blast, or at least so long as iron is melted, by alternate charges of iron and coal. Coal is generally put on first, then iron, and on the top of these the limestone is laid. When all the iron needed for the occasion is melted, the charges are stopped. The blast, however, is urged on, until all the iron has been tapped. The sand bottom of the furnace is made sloping, so as to admit of discharging the last portions of the iron. A well-constructed cupola furnace will melt one ton of iron every hour; some furnaces as much as three tons per hour; small ones, frequently not more than half a ton in an hour. Most furnaces are wider at the bottom than at the top; they therefore work hotter than those with parallel sides, and also have the advantage of lasting longer, as the melted iron, which is apt to cut the fire-brick, does not run down along the brick. The taper to be given to a lining is dependent upon the size of the cupola; a large furnace will bear more taper than a narrow, or small furnace. If different kinds of iron are to be melted in the same heat, a thick layer of fuel is interposed between the various qualities, so as to admit of the extraction of all the iron which was first charged before the second appears at the bottom. In such cases, it is advisable to melt the gray iron, or that iron which is to make soft castings first, and the white or hard iron last. When as much iron is melted as is needed for filling one or more moulds, the clay plug of the tap-hole is pierced by a sharp, steel-pointed bar, and the metal run into pots, which are carried by hand or with a crane, or it is run directly into the mould by means of gutters moulded in the sand of the floor. Between each successive tapping of the iron, the tap-hole is closed, and more iron gathered. Where more iron than the furnace will hold is required for one cast, a portion of it is tapped into a large pot, which process may be carried so far, as to make castings of five or more tons from a small furnace.

18 *

Pots in which iron is carried from the furnaces to the moulds are represented in figures 38 and 39.

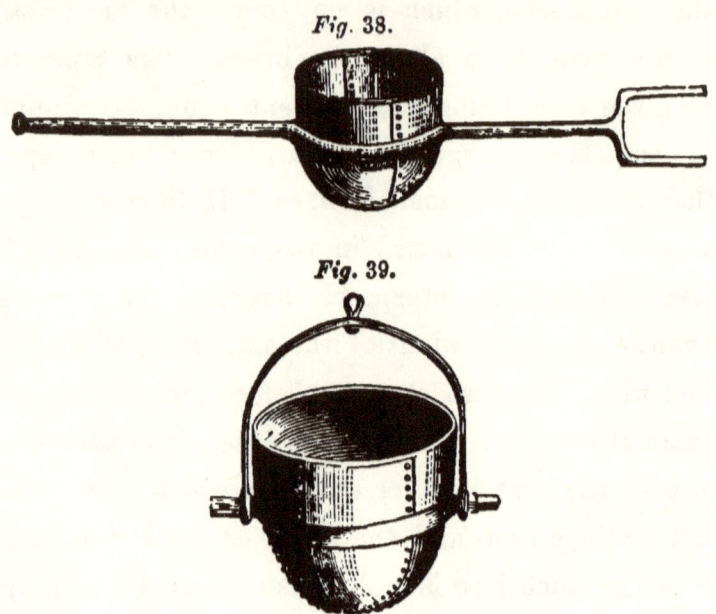

Fig. 38.

Fig. 39.

The first is generally of a capacity sufficient to hold from two to three hundred pounds of melted iron. It can be carried by three or more men; the forked part of the handle is used for tipping the pot, so as to pour the iron gradually into the mould. Figure 39 is designed to be raised by means of a crane, and emptied therefrom into the mould. The cupola or reverberatory at which such a pot is used, as well as the mould, should be within the sweep of the crane. Pots of this description are of various sizes; we find some which will hold five hundred pounds, and others will hold two tons and more. The swivels on such pots are generally strong, and their ends square, with a key-hole to fasten one or two forks to them, for the purpose of tilting the pot and pouring its contents into the mould. These pots are always made of boiler-plate, as it would be dangerous to make them of cast-iron. Before each cast, the pots receive a wash of strong clay-water, to prevent corrosion by the hot iron.

The foregoing are the most important means of melting metal; in the cupola, no metal but iron is melted. Copper, bronze, brass, German-silver, silver, gold, and the alloys of these metals, are either melted in crucibles, or, if large quantities are to be smelted, in the reverberatory furnace. The furnaces, crucibles, and other tools, are essentially the same for other metals as those described for melting iron. Slight variations in the form of melting apparatus are often advised, but there is no essential difference, no alteration in the principle. Fusible metals, such as lead, tin, zinc, antimony, and the alloys of those metals, may be melted in iron pots, kettles, crucibles, and iron ladles, and also in clay crucibles. The heat required to melt these metals is not so high as the melting-heat of iron.

BLAST-MACHINES.

Formerly, cylinder blast-machines were used to supply the cupola with air for combustion, and in some few establishments they are still retained for fanning the furnaces; the impression being, that iron melted by cylinder blast is stronger and less injured than that melted by other blast-machines There is no doubt that the cylinder blast is preferable to the blast generated in machines where water is in contact with the compressed air; in all other respects the impression is erroneous, as there is evidence sufficient to satisfy the most sceptical. In the present case, only, a blast is required for the cupola; in other furnaces it is not needed. To nourish a cupola, no better or more perfect blast can be generated than that made by the fan, or the centrifugal blast-machine. Practice has proved that the fan makes the cheapest blast, and also saves fuel; it has no deteriorating influence upon the iron, provided the quantity of blast sent into the furnace is sufficient to generate a strong heat. In figures 40 and 41, a common fan is represented. It is an iron box, consisting of two cast-iron sides, with a rim of sheet-iron between them. In the centre of the box is a hori-

FOUNDING. 213

zontal shaft, with four fans or wings, which move with great rapidity, drawing in the air at the centres on each side, and driving it towards the periphery,

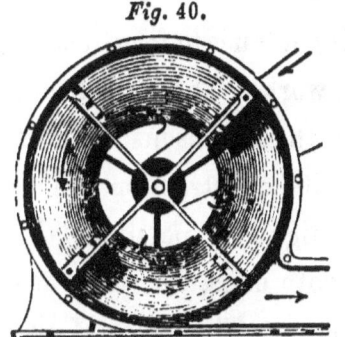

Fig. 40.

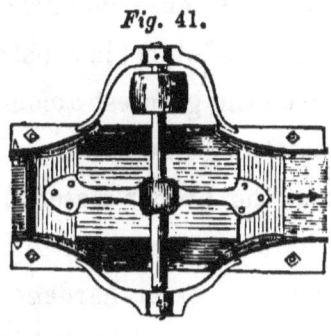

Fig. 41.

thus imparting to the particles of air a momentum, by the centrifugal motion, which presses them against the circumference, and if there is any opening at the circumference, the air will escape with a speed proportionate to that pressure. These fans have been constructed of various sizes and forms; their depth is varied according to the quantity of air to be derived from them; the wings are from four to twenty-four inches wide; eight inches wide is sufficient to supply a well-sized cupola. The diameter is as various as the width of the fan, but it is generally admitted that three feet in diameter is the most profitable and practical size. The wings are often placed in the direction of the diameter, as is shown in the engraving; sometimes in an inclined position to the diameter; and also have been curved in a spiral line, but without any appreciable difference in effect. The latter form of the wings does not cause as much noise as the radial vanes. The chief object in constructing a fan is to form it so as to do the greatest amount of work. The case should be strong and solid, and for these reasons wood is not the proper material for its construction. The shaft and vanes are to be as light as possible; the shaft, of steel, hardened at both ends, where it runs in brass, steel, or cast-iron pans. The vanes of the fan are to be of thin sheet-iron or sheet-copper, and the arms to them of wrought iron. One of the most important conditions of a fan, is the equal weight, and the equal distance from each other of the vanes; and each arm supporting them is to be exactly of the same weight as the other. If these conditions are not complied with, the machine will shake, and soon be out of order. A mere adjustment of the axis, and the vanes attached to it, is not sufficient; it is absolutely necessary, for a good machine, that all the parts around the shaft should be of an equal thickness. In a fan of three feet diameter, the centre openings are generally one foot; in larger fans the openings are larger. Very large apertures will not answer; the air is conducted too quickly to the periphery of the vanes, and there is not sufficient time to impart to the particles the momentum requisite to produce a good effect.

The chief difficulty in constructing a fan is, in the close fitting of the vanes to the sides of the case. The latter cannot be made very straight without incurring much labour, and, on the other hand, it would be very difficult to adjust the axle so perfectly in the centre of the case as not to touch it, which, considering the great speed of the vanes, is almost impossible. It is also easily perceived that the loss in pressure is in the space between the vanes and the cast-iron sides of the case. To diminish this loss, fans are now constructed in which the vanes are covered on both sides with two concentric plane rings, so that the axle with the vanes, forms a hollow drum, open in the centre and at the periphery. The vanes are fastened to these two bottoms or rings, and also to the arms, radiating from the centre. The two bottoms move round with the axis, and parallel with, and close to the sides of the case. In the centre, where the air is drawn in, the case is turned perfectly round, as well as the rim on the centre of the bottoms; both fit closely, but do not touch each other. Where these join there is but from eight to twelve inches diameter, which may be kept tighter than the larger surface and circumference at the vanes. By these means the loss in pressure is greatly diminished, and it is an established fact that these fans require less power, and make stronger blast than fans of other descriptions. Fans of this construction are now most commonly used. The bottoms and vanes in these fans are made of thin sheet copper. The effect of a fan does not depend so much upon its size, as upon its speed and the size of the nozzle. It does not require large vanes to make strong blast; it is sufficient if the surface of each is one-and-a-half times the area of the nozzle, or, if there are more nozzles than one, of the sum of the areas of all the nozzles. More than four vanes in one fan are useless. In the conducting-pipes from the fan to the furnace, there is to be a throttle-valve at each nozzle to shut off the blast at each without disturbing the others. The speed of the axle of a fan is from seven to twelve hundred revolutions per minute. It is driven by a belt and pulley on one side of its axle. To melt a ton of iron in an hour's time, requires about seven hundred cubic feet of air per minute, or, by a three-foot fan, eighteen hundred revolutions, and two three-inch nozzles. Six horses power is needed to drive a fan with the above speed and size of nozzles.

Hot blast has been tried in various instances, but not with such results as to induce a continuance of it. In this instance, hot blast has no other advantage than a small saving of fuel, and as the fuel consumed is not to be considered expensive, the getting up of the apparatus, repair, and disturbances caused by it, amount to more than the gain of fuel.

Drying Stoves are simply brick chambers, one side of which is entirely open. Three sides are formed by a nine or twelve-inch brick wall. In one of the sides is a fire-place, which can be supplied with fuel from the outside of the stove, and may be shut by a close-fitting iron door. In the opposite side of the fire-place is a flue which leads to a chimney; this flue is also low down, almost below the ground. The three sides are covered by a brick arch. The fourth side is provided with iron doors, which open to both sides, and leave the whole fourth side open to any piece of moulding which may be put in. Iron shelves are generally put up along the walls towards the roof, for drying small cores and boxes on. A railroad, which is within the sweep of a crane, leads into the stove, and any heavy mould which is to be dried may be laid upon a car running on this track, and both car and mould are shoved into the stove, the doors closed, and fire put in the furnace. The size of a drying-stove is varied according to the size of the castings commonly made in a foundry. A stove of twelve feet in all directions, and seven feet high, is a good-sized stove. Foundries which make large castings have to be provided with drying-stoves of the proper size. There are frequently more than one drying-stove in a foundry, often as many as five or six, small and large. If there is no occasion for using a large stove, a small one is selected, because it works faster, and with less fuel. In figure 43 a drying-stove is represented.

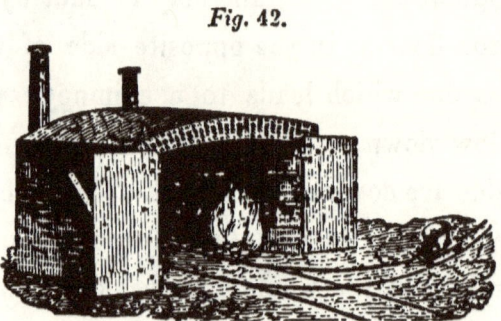

Fig. 42.

GENERAL REMARKS.

Cleansing of castings.—When the metal of a cast is so far cooled as to adhere together, and strong enough to bear removal, the moulds are taken apart and the sand or loam is removed from the casting. Small castings require but a few minutes to cool, while heavier casts take hours and days. A massive casting, such as a forge-hammer of five tons weight, will take twenty-four hours cooling in a green, and forty-eight hours in a dry mould. A bed-plate for the engine of one of the New York line of Atlantic steamers, weighing thirty-five tons, took a week for cooling and the removal from its mould. Heavy castings are chained to a crane and hoisted by it. Very heavy castings require the united strength of two and more cranes. Small castings are removed from their moulds by tongs; one, two, or more persons taking hold of it at the same time, carry it to a place designed for the reception of such hot castings. The excrescences which may happen to have been formed in the partings or core-joints are broken off as soon as the cast is removed to the general deposit of hot castings. The gates are, at the same time, broken off by the moulder; it requires some degree of skill to break a gate off smoth. Gates and accidental excrescences which cannot be removed in the foundry, are chiselled and chipped off in the yard or in the cleansing-shop. Heavy cores, and particularly hard cores, are removed in the foundry before the casting is entirely cold.

Time of casting.—The casting in iron foundries is generally performed in the afternoon after three o'clock, so as to make it the last business of the day. This time is chiefly selected to escape the heat of the hot sand after casting, which will then cool during the night. After casting, the castings are removed. and the moulding-boxes piled in a corner of the building, so as to be handy for the next day's work; the sand, after receiving some water, is shovelled over, mixed, and thrown in heaps, where it remains during the night. If the latter work has been properly performed, the sand will be of a proper and uniform dampness the next morning. Each moulder takes charge of his own sand, and but little practice is required to learn the proper amount of water to be used in damping the sand.

The cleansing of castings is a simple operation in an iron foundry where common castings are made: any workman is fit to trim a coarse casting, or scour it. The first is done by means of chisels or sharp hammers; the latter, with dull, coarse files, which have been used and rejected by machinists. Cast-iron files are also used for the latter purpose. The trimming and cleansing of valuable castings, such as statues or ornaments of art, is not so easily performed. An unskilful workman could undo almost the whole casting, and all the labour spent upon it, by trimming off a channel or gate. This kind of work is done by an artist skilled in the performance of such labour; and, on valuable statues, it is performed by the original designer of the work, at least so far as particular parts, such as the face, or characteristic elements, are concerned. The trimming of fine castings is an art in itself, which requires more explanation than our limited means allow us to give.

The expenses of moulding and casting are very variable. Moulding of common articles of commerce and machinery in iron, is done by the ton, at prices varying from two to twelve dollars per ton, and even at higher prices. Dry-sand moulding is paid higher than moulding in green-sand, and loam-moulding higher than either of them. The moulding of brass, bronze, or other metals, for monuments of art, is of such variety, and so different are the expenses, that no standard price can be assigned to it. The expenses incurred in melting metal are not very great, —the loss in the metal which is melted is greater than the labour and fuel in melting it. In the cupola, twenty-five per cent. of fuel is consumed in melting iron, including all the fuel used in warming the furnace, the drying stoves, and other incidental uses of fuel. Besides fuel, there are two labourers at the cupola, one smelter, and one filler. The reverberatory takes from seventy-five to one hundred pounds of fuel to each hundred pounds of iron, including the heating of the furnace. Exclusive of warming, the reverberatory will take but fifty pounds of fuel. One workman can do the work at the reverberatory, but there are generally two. The melting of iron in the crucible is the most expensive: it consumes from fifty to two hundred pounds of coal to one hundred pounds of iron. The greatest expenses are, however, in the crucibles: a good crucible, well-managed, will not last more than twelve heats, and if each heat is fifty pounds, it will melt six hundred pounds of iron. A crucible of this kind will cost fifty cents; but very few crucibles will melt six hundred pounds, and, on an average, not more than three hundred pounds can be calculated upon.

The loss in iron is invariably from five to six per cent. in every case of the different forms of melting; the reverberatory furnace consuming most iron. Each casting always requires more metal than it will finally contain; this surplus iron, consisting of gates, channels, and false seams, increases the above loss; and as small castings make more scrap iron than large ones, it is obvious that the actual loss will be larger on small casts than on large ones. Machine castings make, on an average, thirty-three per cent. of refuse or scrap in a well-conducted foundry. Commercial articles twenty-five per cent., and large castings less ; very small articles frequently make more scrap than ware. The remelting of these scraps costs fuel, and causes a waste of metal, which increases the expense of melting.

Other metals than iron are generally less expensive in melting, being more fusible; and, as far as copper is concerned, there is but little waste if the copper is pure. Bronze will waste a little ; the waste in volatile metals, as tin and zinc, can be prevented in some measure, if the surface of the melted metal is covered with a mixture of equal parts of potash and soda, mixed with some charcoal powder. To melt and make bronze in the reverberatory, the copper is melted first, and if there is any bronze on hand, in scraps or other forms, it is added as soon as the copper is melted down; after this, the tin is laid near the liquid copper, upon the hearth, and if any zinc or antimony is to be used, it is added last. Before casting bronze, it is to be well stirred by previously heated iron bars. The amount of potash and soda used to protect the metal, is two pounds to one ton of metal; it is added when all the metals are melted and a white scum is visible on the surface of the metal. Bronze metal designed for strong castings, particularly bells, ought to be exposed to the fire in a fluid state for at least eight or ten hours; this will give it a more homogeneous texture and less crystallization. If any zinc is to be added to such an alloy, it is advisable to add it in the form of brass, calculating, of course, the quantity of copper it contains. The relative quantity of the metals forming the alloy can be calculated and mixed according to this arrangement; but the melting operation has an influence upon the strength of the metal. Tin or zinc may be evaporated, and the alloy would not be of the quality intended; the founder, therefore, takes proofs before casting, and if they are not satisfactory, either copper or tin is added to the melted mass. It requires some experience to judge of the quality of an alloy by appearances. Proof is taken by a small iron ladle, the little metal in it is broken after it has cooled, and the form of crystallization and the tenacity of the metal is decisive of the quality of the composition.

Lead, tin, and antimony may be melted in a reverberatory furnace; brass, however, is to be melted in crucibles. Brass is sometimes made by melting copper, and adding, after it is melted, as much zinc as is needed to form the alloy. A cheaper method is to melt a mixture of copper scraps and zinc ore together with some charcoal powder; or, melt both copper and zinc ore together with carbon. In both cases, however, the brass is to be remelted, because the first smelting does not produce strong and pure brass.

APPENDIX.

RECEIPTS AND TABLES.

Alloys of Iron.—All admixtures added to iron make it more fusible than it originally is; these may be metals or metalloids.

Sulphur causes iron to be more fusible if melted together, but this mixture is more liable to corrosion than pure iron. A little sulphur does not injure cast iron, but more than one per cent. makes it brittle when cold. If there is any sulphur in iron when hot, it causes the iron to be brittle.

Carbon is contained in all cast iron from two to six per cent.; it makes the iron fusible; if the amount contained is too large, it renders it brittle. A little carbon makes cast iron brittle and hard. Hard cast iron assumes as beautiful a polish as hardened steel.

Phosphorus makes iron brittle when cold. It imparts a brilliancy and white colour to iron more perfectly than any other matter. Phosphorus makes iron very hard, but renders it liable to corrosion; one-half or one per cent. causes a great alteration in the quality of iron.

Silicon is a constant companion of cast iron; hot-blast iron contains more of it than cold-blast; it also contains more sulphur. and phosphorus if any is present in the ore or coal. Silicon makes iron brittle and hard, and has a similar effect on it as phosphorus.

Arsenic imparts a fine white colour to iron, but makes it brittle.

Chromium causes iron to be as hard as diamond, but it is difficult to make this combination.

Gold combines very readily with iron; it serves as a solder for small iron castings, as breast-pins and similar articles.

Silver does not unite well with iron, but a little may be alloyed with it; it causes iron to be very hard and brittle. The alloy is very liable to corrosion.

Copper, if alloyed with iron, causes it to be brittle when hot, but increases its strength considerably when cold, if the amount of copper is not more than one-fourth of 1 per cent.; more copper than this causes cold-short.

Tin, with iron, makes a hard, but beautiful alloy, which, if nearly half-and-half, assumes a fine white colour, with the hardness and lustre of steel.

Lead combines with iron, but, like silver, in a small proportion; it causes iron to be soft and tough.

Alloys of iron are very little in use at present, but we call attention to such alloys, because the easy method by which, at present, iron is gilded, silvered, or coated with other metals, and also the covering of iron with glass, enamel, and varnish, may, and undoubtedly will lead to the use of iron alloys with greater advantage than the common cast iron is used.

Alloys of Precious Metals.—There are but few which claim our attention. The gold coin of the United States is composed of 90 parts of gold, 2.5 silver, and 7.5 copper; 75 parts of gold, 25 of copper, and often a little silver, is the composition for most trinkets; 66.6 gold, 16.7 silver, and 16.7 copper, forms the solder for gold and iron. Fine silver plate and medals are generally composed of 95 parts silver and 5 copper. Silver solder is 66.6 silver, 30.4 copper, and 3.4 brass.

Alloys of Copper are the most numerous and useful. *Bronze*, or bell-metal, is one of the most beautiful of these alloys.

72 parts copper, $26\frac{1}{2}$ parts tin, and $1\frac{1}{2}$ parts of iron is said to be a superior bell-metal. Iron, tin, and copper do not unite well if each is added separately to the other, but if tin-plate scraps are melted in a crucible together with tin, and then this tin

and iron alloy added to the melted copper, it will unite readily.

Common Bell-metal consists of 100 parts copper and 30 or 40 tin; it is more brittle and of not so good a tone as the other. Another receipt prescribes 78 parts of copper and 22 of tin as a first rate bell-metal. Another highly recommended composition is 80 copper, 10.1 tin, 5.6 zinc, and 4.3 lead. The latter composition is of a good sonorous sound, even if the mould has not been quite dry. The silver bell of Rouen, France, consists of 80 copper, 10 tin, 6 zinc, 4 lead. Too much tin causes the composition to be very brittle. Some bell-founders recommend the addition of a small portion of silver to the composition, but it appears there is no particular necessity for it.

Bronze of great tenacity is composed of 9, 10, or 11 parts of copper to 1 of tin. If this alloy is cast in large masses, it has the peculiarity of separating into parts which contain more or less tin or copper. The tin is generally found on the higher parts of the cast, the copper predominating in the lower parts. This composition, besides being strong, is very hard, and resists abrasion very effectually; it also is very little acted upon by the atmosphere. The ancients used to make their weapons and edged tools of a similar composition,—to which, however, a little phosphorus appears to have been added,—before the invention of steel. If bronze is suddenly cooled, by heating and plunging it in cold water, it becomes less dense and hard, and increases its malleability; but this is not the case in the same degree with all compositions, but the tone of the metal is decidedly impaired, and bells ought never to be cast in damp moulds. Bronze made of the last composition is improved by being tempered, while the tenacity of bell-metal, by the same process, is reduced to one-third of its original strength. The alloy of 80 copper and 20 tin bears tempering best, and increases in strength. The gongs or cymbals, and tamtams of the Chinese, are composed of 80 copper and 20 tin. To give these musical instruments their sonorous property, they are plunged in cold water after being cast; a reheating to ignition, however, is to precede the refrigeration. After this latter process, which deprives the metal of almost all its sound, it is tempered, and very slowly cooled, which imparts to it the capacity of emitting that peculiarly powerful sound.

Bronze for Statues is of a great variety of composition. We also find alloys for this purpose composed like bell-metal, and also of almost pure copper

Modern statues are composed of a composition of 80 copper and 20 tin. The present state of the art of making valuable bronze castings is, however, so imperfect, that our age cannot be considered competent to give a standard of metal compositions for that purpose. The French artists, in the first part of this century, were so ignorant in this peculiar art, that some parts of the Vendome column are an alloy of 94 copper and 6 tin, while other parts contained but $\frac{1}{5}$ of alloy to $99\frac{4}{5}$ of copper. These defects caused bad castings, so that the chisellers had to cut off seventy tons of protuberances on this one monument. At the time of Louis XIV., a period when the art of casting statues was more cultivated in France, statues were cast of an alloy consisting of 91.3 copper, 1 to 2 tin, 5 to 6 zinc, and 1 to 1.5 lead. The statue of Louis XV. is cast of copper 82.4, zinc 10.3, tin 4, and lead 3.2.

The Bronze of the *Ancient Greeks* consisted chiefly of copper and tin, but was frequently alloyed with gold, silver, lead, zinc, and arsenic. The Greeks not only made statues, tripods, lamps, and other articles of art of bronze, but also their weapons, shields, coin, nails, kitchen utensils, and chirurgical instruments. The ancients understood the art of hardening and tempering bronze to perfection, so that the want of steel was not so severely felt as we may be inclined to believe at the present time.

The Ancient Mexicans—Aztecs—understood the art of converting bronze into edged instruments in a high degree. To small castings, an addition of iron, in the form of tin-plate scraps, appears to be advantageous: large articles are liable to crystallize by the addition of that metal.

Speculum Metal is generally composed of $66\frac{1}{3}$ copper and $33\frac{2}{3}$ tin, it is white, and has a brilliant lustre, and is susceptible of a high polish. An ancient mirror was found to consist of 62 copper, 32 tin, and 6 lead. In France, 2 parts of copper and 1 part of tin are used, which are melted separately in crucibles, and mixed just before casting. The addition of a little arsenic, one or two per cent., makes the metal more compact, and gives it a greater lustre and hardness, but renders it liable to be tarnished by the air. *The speculum metal* of Lord Rosse's large telescope is composed of 126.4 copper and 58.9 tin. This alloy is of a brilliant white lustre, and has a specific gravity of 8.811; it is nearly as hard as steel, and as brittle as sealing-wax. The speculum is cast 6 feet in diameter and $5\frac{1}{4}$ inches thick, and weighs upwards of three tons. The casting of this mirror was an interesting process. After repeated failures and experiments, a mould was made whose bottom consisted of a wrought-iron ring, packed full of hoop-iron laid edgeways, so close that air, but no metal, could escape through the crevices. This bottom was turned convex on a turning-lathe, true to the concavity of the speculum; it was then placed upon a level floor and enclosed by a sand-dam, left open from above. The metal was melted in cast-iron crucibles, because wrought iron or clay would have injured the alloy. The cast was carried while red hot into the annealing oven, which was previously heated to a red heat, and left there sixteen weeks to cool.

Bronze for Medals generally contains least tin. 100 copper with 4.17 tin has been proposed, but this alloy is so hard, that it has been found necessary to cast the coin. Bronze medals are, however, stamped when composed of 92 copper and 8 tin, a little zinc being added in a form of brass.

Bronze in imitation of Gold, consists of 90.5 copper, 6.5 tin, and 3 zinc.

If bronze is to be gilt, it should be of easy fusion, and take perfect impressions of the mould. A combination of copper, tin, zinc, and lead, as previously noticed for statues, is the best in this case. An alloy which is said to possess the best properties for being gilt, was composed of 82.25 copper, 17.48 zinc, .23 tin, .02 lead. An alloy for gilding is to be compact and of close grain. It absorbs gold and mercury in proportion to its porosity.

Brass is a composition of copper and zinc; 2 parts of copper and 1 of zinc—or more correctly $63\frac{1}{2}$ copper and 32.3 zinc—form common brass. Two parts of brass and one of zinc form hard solder; to this a little tin may be added. If the solder is to be tough, as for pipes or kettles, which are to be drawn or beaten, but $\frac{2}{3}$ of zinc are to be added to 2 of brass. *Button-brass* consists of 8 parts of brass and 5 of zinc. *Red-brass* or tombak is made of 8 or 10 copper, and 1 zinc, or, as in some German works, of 11 copper and 2 zinc. *Princes metal*, Similor, Nurnberg gold, or Manheim gold, are different compositions, varying between 3 copper and 1 zinc, and 2 copper and 1 zinc. These elements are separately melted, and mixed together by constant stirring. Brass containing a little lead, from one to two per cent., is more easily turned than common brass, but is more brittle. Brass which is best adapted for hammering consists of 70 copper and 30 zinc. Tempering and sudden refrigeration has a similar effect on brass as upon bronze; the first renders it hard and more tenacious, and the latter soft. A little zinc makes a reddish brass, and imparts a golden hue; larger quantities make it a greenish yellow, and more than fifty per cent. of zinc causes brass to be of a bluish gray colour. *Brass for ship nails* consists of 10 copper, 8 zinc, and 1 iron. Brass for pans and steps to run machine shafts in, is to contain less zinc than common brass; an addition of bronze to brass increases its applicability for such purposes. It is said that 16 copper, 1 zinc, and 7 platinum is almost equal to gold. If melted red-brass is stirred with an iron or steel rod, so as to impart a little iron to it, its strength is sensibly augmented. The variety of brass compositions is so numerous, as to make it impossible to note all the known compounds. In the above, the most useful alloys are enumerated.

German-silver, Argentan, or Chinese Packfong, is one of the most valuable alloys; it nearly combines the durability of silver and the utility of iron, steel, and copper. Common German-silver is composed of 60 copper, 25 zinc, and 15 nickel. A better quality is 50 copper, 25 zinc, and 25 nickel. *Chinese packfong* consists of 55 copper, 17 zinc, 23 nickel, and 3 iron. A highly sonorous, tenacious *Argentan*, which can be hammered and rolled, resembling silver more than any other compound, is composed of

10.4 copper, 25.4 zinc, 31.5 nickel, 2.6 iron. At present, a fine argentan, and one the best qualified to be plated with silver by the galvanic process, is made of 62 copper, 19 zinc, 13 nickel, and 4 to 5 cobalt and iron. This argentan is very close, strong, and cheap, and may be covered by one or two per cent. of silver, forming a good fine plate. A very tenacious, ductile, and hard argentan may be made of 57.4 copper, 25 zinc, 13 nickel, and 9 iron. This alloy can be substituted for steel in many of the common uses of steel, particularly where corroding influences upon steel are strong, because this alloy is not affected by atmospheric air. *Electron*, a fine quality of argentan, is composed of 8 copper, 4 nickel, and 3.5 zinc. *Solder for German silver* is made by adding 4 parts of zinc to this composition, then laminate and pound it to a coarse powder.

Before we part with copper alloys it will be proper to allude to some combinations of copper with other matters which are useful to know. *Copper and platinum* form a yellow alloy hardly distinguishable from gold. *Copper and silver* do not form any distinguished amalgam; the addition of a little arsenic to such an alloy makes it whiter and more like silver. A little copper and antimony make a fine rose-coloured alloy; if the copper is increased, it assumes a darker hue; equal quantities make a violet compound, and more copper increases the dark shade This alloy is brittle in all proportions; 90 parts of copper, 5 antimony, and 5 zinc, are used for plumber-blocks, and pans and steps for steel and iron gudgeons to run in. *Carbon* makes copper very brittle. *Phosphorus* makes copper as hard as steel, so that it can be used for knives and edge-tools; it, however, renders copper more liable to corrosion. The appearance of this compound when newly polished is like pure copper, but it is very soon covered or tarnished with a greenish-black covering. This greenish black being the colour of ancient weapons, renders it probable that the ancients hardened their copper or bronze tools by means of phosphorus. *Copper and arsenic* form a bright white alloy, which is used for candlesticks, buttons, dials, and similar articles, but as this compound is easily soluble and highly poisonous, it cannot be used where food is brought in contact with it. This alloy is made by melting copper scraps and white arsenic—arsenious acid—in a crucible, covering it with a layer of common salt. It has almost the colour and density of pure silver, but is very liable to corrosion.

Lead and its alloys are very extensively used; the alloys are usually harder and less tough than lead. A *small amount of arsenic* is added to the lead to make shot; arsenic is more fusible and more brittle than lead; for fine shot, three pounds of arsenic, and for coarse eight pounds, to one thousand pounds of lead, are generally used. To alloy lead with arsenic, nothing more is needed than to melt white arsenic together with metallic lead; half the weight of the arsenic employed will be absorbed by the lead. 5 *lead and* 1 *antimony*, to which frequently a little zinc and bismuth are added, forms type metal. A good French type metal is said to consist of 2 lead, 1 antimony, and 1 copper. Common type metal is 80 lead and 20 antimony; a more fusible stereotype metal is 77 lead, 15 antimony, and 8 bismuth. Some stereotype founders add tin to the above, that is, add to lead, antimony, and bismuth, tin; or leave the bismuth out and supply its place by tin. If much tin is used it renders the metal rather soft, but fusible and fit for fine impressions. A superior alloy is said to consist of 9 lead, 2 antimony, and 1 bismuth. To alloy lead with these metals, the lead is first melted, and the other metals added to the fluid lead. *Fusible metal* may be compounded of various degrees of fusibility; 31 lead, 19 tin, and 50 bismuth may be fused at 203°. An alloy which fuses at 149°, and which is used for plugging teeth, consists of 28.5

lead, 45.5 bismuth, 17 tin, and 9 mercury. 8 of bismuth, 5 of lead, and 3 of tin, will melt at the boiling heat of water, or 212°. Bismuth makes lead stronger if the amount of bismuth does not exceed that of the lead; two parts of bismuth and three parts of lead is said to be ten times stronger than lead, and as the durability of bismuth is equal to lead, it forms a good alloy for making pipes and wire.

Tin forms a range of very useful alloys. *Tin and lead* melt together in all proportions. Most of the tin vessels which are called pure tin are alloyed with lead. Soft solder is 33 tin to 67 lead, and in all proportions from that to 67 tin to 33 lead; half-and-half is common soft solder. *Plate pewter* is composed of 89 tin, 2 bismuth, 7 antimony, and 2 copper. *Queen's metal*, of 75 tin, 9 lead, 8 bismuth, 8 antimony. *Britannia metal*, of 89 tin, 2 copper, 6 antimony, 2 brass, and 1 iron. *Common pewter*, or German tin, is composed of 4 tin and 1 lead. The best plate pewter is 100 tin, 8 antimony, 2 bismuth, and 2 copper. *Music metal* is 80 tin and 20 antimony. *Spurious silver leaf* is 50 tin and 50 zinc. Antifriction metal is a variable compound of lead, antimony, tin, and copper. *Organ pipes* are made of a composition of 9 tin and 1 lead; these proportions are varied by different artists. 29 tin and 19 lead form a fusible compound, of which imitations of diamonds and precious brilliants are made. To make such imitations, a glass rod is ground at one end in the form which is to be represented, whether a brilliant or rose-diamond. The melted metal is skimmed by a paper card, and the ground facetted end of the glass rod dipped in the clear metal; on withdrawing the rod a thin film of metal will adhere to the cold rod, which, when taken off, will show a hollow capsule having the lustre of a diamond. We find such diamonds at present used to make sign-boards in show windows. This metal forms excellent reflectors, which may easily be made by dipping a round bottle or the bottom of a retort in the metal; but the metal is tarnished by anything coming in contact with it. 1 part tin, 1 lead, 2 bismuth, and 10 mercury is very fusible; with this compound glass pipes and glass globes are coated with a thin film, by placing some of this metal in the article to be coated, and allowing it to flow round, thus giving it the brilliancy of silver. *Tin foil*, if designed for mirrors, is pure tin, but common tin-foil is lead and tin—often tin, zinc, and lead; it has so great a variety of composition, that no standard can be assigned it. Tin-foil is made either by hammering or rolling, but most of it is made by casting the hot metal over an inclined plane, made of a frame covered with cotton or linen canvas. It requires some skill to perform the latter operation.

Zinc, alloyed with other metals, has already been enumerated. In its pure state it forms fine sharp castings, good for ornamental purposes; but as these castings have no strength, they are not much used for other purposes. A composition of lead and zinc is used for patterns, but with little advantage; it is soft and flexible, and the patterns soon lose their shape.

BRONZING.

When bronze metal has been exposed to the atmospheric air for a long time, it assumes a dark green colour. This colour, a rich hue, may be imitated by chemical agencies, or by paint. Bronze metal, after being cleaned, is bronzed by being painted or immersed in a solution of two parts of verdigris and one part of sal-ammonia, dissolved in vinegar, boiled and filtered, and used very dilute. It is left in this solution or brushed over until the desired hue is obtained. *The colour of antique bronze* is obtained by painting the bronze cast with a solution of one part of sal-ammonia, three parts cream of tartar, six parts of common salt, the whole dissolved in twelve parts of hot water; with this are to be mixed eight parts of a solution of nitrate of copper. This solution should be laid on in a damp place. The first mixture will give a more reddish dark green colour to bronze than the latter. Different tints may be imparted to bronze and brass, from red to bright yellow, and from dark to light green. *Boiling bronze* in muriatic acid will give it a red colour; and soaking it in ammonia renders it whiter than it already is. Bronze painted with a thin solution of equal parts of sal-ammonia and oxalate of potash, in a warm room, or in the heat of the sun, gives it a fine green colour, particularly if rubbed with it. If a dark blackish bronze colour is required, the foregoing solution is laid on in a room where some liver of sulphur—sulphuret of potassium—is dissolved in water, and set out in flat dishes to generate sulphuretted hydrogen, which will cause a uniform blackish brown colour on bronze or brass. The foregoing receipts answer for brass as well as bronze. When the desired colour is obtained, the object is washed in clean water, dried, and then rubbed with a brush and wax. The bronze for the latter operation is heated, but not so much so as to burn the wax.

Bronze colour is imparted to other castings besides brass and bronze, by paint. Cast iron may be bronzed by dipping it in a thin solution of sulphate of copper, or muriate of copper, and when sufficiently covered with copper, it is washed and painted with oil varnish. All objects to be bronzed may, however, be painted of any colour, either a shade of green, from the faintest to an almost black green, or of a blue or bluish green. The paint cover should be coated with pure varnish, and when that is nearly dry, a metallic powder is dusted over it by a dusting-bag, or rubbed on by the fingers, a linen pad, or a paint-brush. The metallic powder is generally mosaic gold, which is made of almost every shade, and is of great beauty; or it may be copper in powder, gold leaf, silver leaf, and similar articles; dry paint of a convenient shade may also be used. The highest parts of the articles are generally bronzed so as to appear as if rubbed and worn by use. Over the whole of these, a last coating of spirit varnish is laid on.

The gilding of bronze and brass castings is performed, in the dry way, by making the surface perfectly smooth, then brushing it over with an amalgam of gold, and dissipating the mercury by heat, which leaves a durable film of gold over the surface. This surface may be burnished or deadened. The amalgam is made by heating one part of gold, in thin laminæ, in a crucible, and when it becomes faintly red, pour over it eight parts of mercury, pour the combined gold and mercury into cold water, and squeeze the surplus mercury out. The amalgam is then enclosed in canvas or chamois leather, and some more mercury pressed out; the remainder will contain one part of gold to two parts of mercury. This amalgam is rubbed over the objects to be gilded: it may be had in its true composition from the gold mines of Virginia, and of the best quality from North Carolina. It is advisable to brush the brass over with a thin solution of nitrate of mercury and some free nitric acid, as this facilitates the adherence of the amalgam. The gilt and burnished articles may be coloured by a simple process to any shade from a bright and crimson red to a violet and deep blue, by being submerged in a bath of caustic potash in which some metallic oxide is dissolved, but, as a galvanic process is to be applied here, it is beyond our province to describe it. There are other methods of gilding which, for the same reason, must be excluded.

Iron may be gilded by brushing it over with a solution of gold in sulphuric ether. The iron is to be bright and polished, and the gold rubbed on by the burnisher. This is not very durable gilding.

Tinning of brass, bronze, and copper, is done by washing the surface of the cast with very diluted sulphuric acid, after which, wash in water, and scour with sand. The object is then heated to the melting point of tin, and the tin, having been previously melted, is rubbed over the surface by means of a damp rag or piece of oakum, first covered with rosin, to protect the tin against oxydation. Cast iron must be turned or filed, so as to offer a clean surface, before it can be tinned. A solution of tin, as muriate of tin, mixed with an equal part of sal-ammonia, if brushed over the metal, will highly facilitate the operation of tinning. A more convenient mode of tinning than the above, is to plunge the object to be tinned in a solution of tin and caustic potash, which solution is to be as hot as it can possibly be made. Such a solution of tin is made by dissolving oxyde of tin—putty of tin— in potash ley, adding to the saturated solution some metallic tin, in the form of filings or shavings of tin. A few minutes are sufficient to cover brass or copper with tin.

Zinking of copper or bronze may be done by exposing the objects to the fumes of zinc. On copper castings, it is often desirable to have some parts of a golden or yellow hue, which may be done by

21 *

exposing those parts to the fumes of zinc. A very perfect coating of zinc may be obtained by placing the objects, well cleaned, in a solution of chloride of zinc, in which a surplus of metallic zinc is present. Chloride of zinc is made by dissolving zinc in muriatic acid, always having so much zinc in the acid, that some of it will remain undissolved. Zinc dissolved in sal-ammonia is as efficacious as the foregoing.

Glazing of metal castings, or coating with enamel, is very little practised, and will hardly ever amount to a lucrative operation. Iron to be coated with an enamel is first well cleansed by means of acid and scouring with sand. It is then uniformly covered with the enamel, which has been previously prepared or melted, finely ground, and mixed with water for the purpose of laying it on. This operation is very little practised, as it is very expensive and the product is not durable. It has been, and still is used for covering the interior of cooking utensils to prevent their cooking black. A better means to accomplish this object in a cheaper way, is the application of cast iron, which contains a little phosphorus, and not too much carbon, as has been previously remarked. More recently, a new invention, that of covering iron with transparent glass, and also with coloured glass, has made its appearance in England. Serious doubts, however, may be entertained as to its ultimate success. Iron coated with enamel or glass can never be brought to a successful competition with porcelain either in beauty or price.

Blackening of iron casts is either done with black-lead, moistened with alcohol, or, in many instances, with spirits of turpentine. This is laid on by a brush, and rubbed until the blackening is dry, and assumes a metallic lustre. This is the blackening used for stoves. If the object to be blackened is a little warm, the operation works better and much more quickly.

Fine ornamental castings are heated to the blue annealing heat, and then covered with black copal varnish, and dried at the same degree of heat. The heat takes most of the gloss of the varnish off. The copal varnish is then blackened by an admixture of finely rubbed lamp-black, or printers' ink, or, still better, by finely ground pure bone-black. Larger castings are blackened with common black paint. A rich lead-colour may be imparted to castings by an oil paint, made of fine litharge gently heated in an iron pan, and, when hot, some flour of sulphur finely and uniformly sprinkled over it under constant stirring. The resulting sulphuret of lead assumes a rich lead-colour, which is not altered by oil or the atmosphere.

Grinding of cast iron is resorted to where any smooth, polished surface is required. It is done on large, fast-revolving sand or grind stones. Cast iron is generally hard on its surface and sandy, so that it would require too much labour to file it, besides wearing out too many files. Machine castings are planed or turned by proper machinery.

Malleable cast iron, an article now very much in use for carriage and harness furniture, and various other purposes, is made of the best kind of No. 2 charcoal pig. Where the foundry scraps are of a good quality of iron, they are preferable. A good article may be made by mixing No. 2 and No. 3 iron. Any pig iron which makes good bar iron will make malleable iron. Most of the malleable iron is cast from the cupola, but the crucible makes better castings of the same material. The cast articles are tempered in an iron cylinder, and imbedded in fine fresh river sand, or finely pounded iron ore, or black manganese, or a mixture of the whole of these materials. An exposure of the hardest cast iron, if pure, from twenty-four to thirty-six hours to the fire, will render it malleable to a certain degree. When tempered, the articles are put in a revolving iron barrel together with some sand, to be cleaned and polished, to a certain extent, by rubbing one against the other. This malleable iron is particularly qualified for being tinned, or plated with brass or silver. For the silvering of iron, a process has been lately recommended which appears to be valuable: it is performed by means of galvanism. The iron article, well cleaned and freed of all oil and grease, is immersed in a solution of silver, and connected with the zinc pole of a galvanic battery; the copper pole is connected with a platinum plate placed in the solution at some distance from the cast iron. The silver solution consists of cyanide of silver. It is made by putting cyanide of potassium in a well-corked vessel, together with freshly prepared chloride of silver; the whole is then covered with water and violently shaken. It is advisable to use an excess of chloride of silver, and if a little remains undissolved, add a few pieces of cyanide of potassium. A little chloride of silver ought to remain after all the cyanide is saturated. This solution is filtered, to render it perfectly clear, and is then ready for use. It is said that a few minutes' time is sufficient to cover a large surface of iron with silver.

TABLE I.

Weight of a lineal foot of cast-iron pipes in pounds.

Diameter of bore in inches.	Thickness of metal in inches.							
	3/8	1/2	5/8	3/4	7/8	1	1 1/8	1 1/4
2	8.8	12.3	16.1	20.3				
2 1/2	10.6	14.7	19.2	23.9				
3	12.4	17.2	22.2	27.6	33.3	39.3	45.6	
		19.6	25.3	31.3	37.6	44.2	51.1	
3 1/2	14.2	22.1	28.4	35.0	41.9	49.1	56.0	64.4
4	16.1	24.5	31.4	38.7	46.2	54.0	62.1	70.6
4 1/2	18.0	27.0	34.5	42.3	50.5	58.9	67.6	76.7
5	19.8	29.5	37.6	46.0	54.8	63.8	73.2	82.8
5 1/2	21.6	31.9	40.7	49.7	59.1	68.7	78.7	88.8
6	23.5	34.4	43.7	53.4	63.4	73.4	84.2	95.1
6 1/2	25.3	36.8	46.8	56.8	67.7	78.5	89.7	101.2
7	27.3							
7 1/2	29.0	39.1	49.9	60.7	72.0	83.5	95.3	107.4
8	30.8	41.7	52.9	64.4	76.2	88.4	100.8	113.5
8 1/2	32.9	44.4	56.2	68.3	80.8	93.5	106.5	119.9
9	34.5	46.6	59.1	71.8	84.8	98.2	111.8	125.8
9 1/2	36.3	49.1	62.1	75.5	89.1	103.1	117.4	131.9
10	38.2	51.5	65.2	79.2	93.4	108.0	122.8	138.1
10 1/2		54.0	68.2	82.8	97.7	112.9	128.4	144.2
11		56.4	71.3	86.5	102.0	117.8	133.9	150.3
11 1/2		58.9	74.3	90.1	106.3	122.7	139.4	156.4
12		61.3	77.4	93.6	110.6	127.6	145.0	162.6
13			82.7	101.2	118.2	137.4	154.1	173.5
14			89.3	108.2	126.5	146.2	165.3	185.2
15			95.2	115.7	135.3	156.2	176.2	198.1
16				123.3	143.1	166.1	187.5	211.3
17				130.2	152.5	178.5	198.2	223.4
18				137.0	161.2	185.3	209.1	235.6
19					169.2	195.7	222.3	247.1
20					178.1	205.2	233.2	259.0

N. B. The two flanges of a pipe are considered equal to the weight of one foot in length.

RECEIPTS AND TABLES. 951

TABLE II.
Dimensions of cylindrical columns of cast iron to sustain a given load with safety.

Diameter in inches.	Height in feet.										
	4	6	8	10	12	14	16	18	20	22	24
2	72	60	49	40	32	26	22	18	15	13	11
3	178	163	145	128	111	97	84	73	64	56	49
4	326	310	288	266	242	220	198	178	160	144	130
5	522	501	479	452	427	394	365	337	310	285	262
6	607	592	573	550	525	497	469	440	413	386	360
8	1333	1315	1289	1259	1224	1185	1142	1097	1052	1005	959

(Weight in cwts.)

N. B. If the columns are hollow, the area to the given diameter is to be converted into the ring, or the difference of the outer and inner diameters multiplied by ⅔, because hollow cast-iron columns are stronger than solid ones in that proportion.

TABLE III.
Showing the tenacities, and resistances to compression, of various simple metals and alloys.

METALS AND ALLOYS.	Tenacity. A bar of one inch square section, will be torn asunder by	Resistance to Compression. One square inch will be crushed by	Resistance to Torsion.
	Pounds	Pounds	
Cast Iron	15,000 to 30,000	86,000 to 100,000	9.0
Copper, Wrought	33,000		4.3
Malleable Iron	56,000 to 70,000		10.0
Lead	1824		1.0
Steel	120,000 to 150,000	200,000 to 250,000	16 to 19
Tin	5000		1.4
Zinc	9000		
Common Brass	17,900	10,300	4.6
Swedish Copper 6 parts, Malacca Tin 1 part	64,000		5.0
Chili Copper 6 parts, Malacca Tin 1 part	60,000		
Common Block Tin 4 parts, Lead 1 part, Zinc 1 part	13,000		
Common Block Tin 3 parts, Lead 1 part	10,200		
Common Block Tin 3 parts, Zinc 1 part	10,000		
Lead 1 part, Zinc 1 part	4500		

TABLE IV.

Specific gravities of metals and alloys. Water 1000.

METALS AND ALLOYS.	Specific Gravity.	The weight of a cubic inch is in pounds	Number of cubic inches in one pound.	The weight of a cubic foot is in pounds	Melting point in degrees.
Platinum	19.500			1208	
Gold	19.258		1.435	1203	2010°
Mercury	13.500		2.038	843	
Lead	11.352	.4103	2.435	708	612°
Silver	10.474		2.638	652	1873°
Bismuth	9.823		2.814	613	476°
Copper, Cast	8.788	.3185	3.146	550	1996°
" Wrought	8.910	.3225	3.103	555	
Iron, Cast	7.264	.2630	3.806	450	2780°
Steel	7.816		3.530	489	
Tin, Cast	7.291	.2636	3.790	456	442°
Zinc, Cast	7.190	.2600	3.845	449	773°
Gold 90, Silver 2.5, Copper 7.5	17.40				
Gold 66.6, Silver, 16.7, Copper 16.7 (Solder for Gold.)	12.40				
Zinc 10.0, Silver, 66.6, Copper 23.4 (Solder for Silver.)	9.84				
Bronze	8.48 to 8.94			537	
German Silver	8.48 to 8.57				
Brass	8.4 to 8.5		3.533	537	1900°
Type Metal	9.854			615	
Soft Solder	9.55				
Music Metal	7.1				
Water	1.000			62.5	

THE END.

CATALOGUE
OF
PRACTICAL AND SCIENTIFIC BOOKS,
PUBLISHED BY
HENRY CAREY BAIRD & CO.,
Industrial Publishers and Booksellers,
NO. 810 WALNUT STREET,
PHILADELPHIA.

☞ Any of the Books comprised in this Catalogue will be sent by mail, free of postage, at the publication price.

☞ A Descriptive Catalogue, 96 pages, 8vo., will be sent, free of postage, to any one who will furnish the publisher with his address.

ARLOT.—A Complete Guide for Coach Painters.
Translated from the French of M. ARLOT, Coach Painter; for eleven years Foreman of Painting to M. Eherler, Coach Maker, Paris. By A. A. FESQUET, Chemist and Engineer. To which is added an Appendix, containing Information respecting the Materials and the Practice of Coach and Car Painting and Varnishing in the United States and Great Britain. 12mo. $1.25

ARMENGAUD, AMOROUX, and JOHNSON.—The Practical Draughtsman's Book of Industrial Design, and Machinist's and Engineer's Drawing Companion:
Forming a Complete Course of Mechanical Engineering and Architectural Drawing. From the French of M. Armengaud the elder, Prof. of Design in the Conservatoire of Arts and Industry, Paris, and MM. Armengaud the younger, and Amoroux, Civil Engineers. Rewritten and arranged with additional matter and plates, selections from and examples of the most useful and generally employed mechanism of the day. By WILLIAM JOHNSON, Assoc. Inst. C. E., Editor of "The Practical Mechanic's Journal." Illustrated by 50 folio steel plates, and 50 wood-cuts. A new edition, 4to. $10.00

ARROWSMITH.—Paper-Hanger's Companion:
A Treatise in which the Practical Operations of the Trade are Systematically laid down: with Copious Directions Preparatory to Papering; Preventives against the Effect of Damp on Walls; the Various Cements and Pastes Adapted to the Several Purposes of the Trade; Observations and Directions for the Panelling and Ornamenting of Rooms, etc. By JAMES ARROWSMITH, Author of "Analysis of Drapery," etc. 12mo., cloth. $1.25

ASHTON.—The Theory and Practice of the Art of Designing Fancy Cotton and Woollen Cloths from Sample:
Giving full Instructions for Reducing Drafts, as well as the Methods of Spooling and Making out Harness for Cross Drafts, and Finding any Required Reed, with Calculations and Tables of Yarn. By FREDERICK T. ASHTON, Designer, West Pittsfield, Mass. With 52 Illustrations. One volume, 4to. $10.00

BAIRD.—Letters on the Crisis, the Currency and the Credit System.
By HENRY CAREY BAIRD. Pamphlet. 05

BAIRD.—Protection of Home Labor and Home Productions necessary to the Prosperity of the American Farmer.
By HENRY CAREY BAIRD. 8vo., paper. 10

BAIRD.—Some of the Fallacies of British Free-Trade Revenue Reform.
Two Letters to Arthur Latham Perry, Professor of History and Political Economy in Williams College. By HENRY CAREY BAIRD. Pamphlet. 05

BAIRD.—The Rights of American Producers, and the Wrongs of British Free-Trade Revenue Reform.
By HENRY CAREY BAIRD. Pamphlet. 05

BAIRD.—Standard Wages Computing Tables:
An Improvement in all former Methods of Computation, so arranged that wages for days, hours, or fractions of hours, at a specified rate per day or hour, may be ascertained at a glance. By T. SPANGLER BAIRD. Oblong folio. $5.00

BAIRD.—The American Cotton Spinner, and Manager's and Carder's Guide:
A Practical Treatise on Cotton Spinning; giving the Dimensions and Speed of Machinery, Draught and Twist Calculations, etc.; with notices of recent Improvements: together with Rules and Examples for making changes in the sizes and numbers of Roving and Yarn. Compiled from the papers of the late ROBERT H. BAIRD. 12mo. $1.50

BAKER.—Long-Span Railway Bridges :
Comprising Investigations of the Comparative Theoretical and Practical Advantages of the various Adopted or Proposed Type Systems of Construction; with numerous Formulæ and Tables. By B. BAKER. 12mo. $2.00

BAUERMAN.—A Treatise on the Metallurgy of Iron :
Containing Outlines of the History of Iron Manufacture, Methods of Assay, and Analysis of Iron Ores, Processes of Manufacture of Iron and Steel, etc., etc. By H. BAUERMAN, F. G. S., Associate of the Royal School of Mines. First American Edition, Revised and Enlarged. With an Appendix on the Martin Process for Making Steel, from the Report of ABRAM S. HEWITT, U. S. Commissioner to the Universal Exposition at Paris, 1867. Illustrated. 12mo. . $2.00

BEANS.—A Treatise on Railway Curves and the Location of Railways.
By E. W. BEANS, C. E. Illustrated. 12mo. Tucks. . . $1.50

BELL.—Carpentry Made Easy :
Or, The Science and Art of Framing on a New and Improved System. With Specific Instructions for Building Balloon Frames, Barn Frames, Mill Frames, Warehouses, Church Spires, etc. Comprising also a System of Bridge Building, with Bills, Estimates of Cost, and valuable Tables. Illustrated by 38 plates, comprising nearly 200 figures. By WILLIAM E. BELL, Architect and Practical Builder. 8vo. . $5.00

BELL.—Chemical Phenomena of Iron Smelting :
An Experimental and Practical Examination of the Circumstances which determine the Capacity of the Blast Furnace, the Temperature of the Air, and the proper Condition of the Materials to be operated upon. By I. LOWTHIAN BELL. Illustrated. 8vo. . . $6.00

BEMROSE.—Manual of Wood Carving :
With Practical Illustrations for Learners of the Art, and Original and Selected Designs. By WILLIAM BEMROSE, Jr. With an Introduction by LLEWELLYN JEWITT, F. S. A., etc. With 128 Illustrations. 4to., cloth. $3.00

BICKNELL.—Village Builder, and Supplement :
Elevations and Plans for Cottages, Villas, Suburban Residences, Farm Houses, Stables and Carriage Houses, Store Fronts, School Houses, Churches, Court Houses, and a model Jail; also, Exterior and Interior details for Public and Private Buildings, with approved Forms of Contracts and Specifications, including Prices of Building Materials and Labor at Boston, Mass., and St. Louis, Mo. Containing 75 plates drawn to scale; showing the style and cost of building in different sections of the country, being an original work comprising the designs of twenty leading architects, representing the New England, Middle, Western, and Southwestern States. 4to. . $12.00

BLENKARN.—Practical Specifications of Works executed in Architecture, Civil and Mechanical Engineering, and in Road Making and Sewering:
To which are added a series of practically useful Agreements and Reports. By JOHN BLENKARN. Illustrated by 15 large folding plates. 8vo. $9.00

BLINN.—A Practical Workshop Companion for Tin, Sheet-Iron, and Copperplate Workers:
Containing Rules for describing various kinds of Patterns used by Tin, Sheet-Iron, and Copper-plate Workers; Practical Geometry; Mensuration of Surfaces and Solids; Tables of the Weights of Metals, Lead Pipe, etc.; Tables of Areas and Circumferences of Circles; Japan, Varnishes, Lackers, Cements, Compositions, etc., etc. By LEROY J. BLINN, Master Mechanic. With over 100 Illustrations. 12mo. $2.50

BOOTH.—Marble Worker's Manual:
Containing Practical Information respecting Marbles in general, their Cutting, Working, and Polishing; Veneering of Marble; Mosaics; Composition and Use of Artificial Marble, Stuccos, Cements, Receipts, Secrets, etc., etc. Translated from the French by M. L. BOOTH. With an Appendix concerning American Marbles. 12mo., cloth. $1.50

BOOTH AND MORFIT.—The Encyclopedia of Chemistry, Practical and Theoretical:
Embracing its application to the Arts, Metallurgy, Mineralogy, Geology, Medicine, and Pharmacy. By JAMES C. BOOTH, Melter and Refiner in the United States Mint, Professor of Applied Chemistry in the Franklin Institute, etc., assisted by CAMPBELL MORFIT, author of "Chemical Manipulations," etc. Seventh edition. Royal 8vo., 978 pages, with numerous wood-cuts and other illustrations. . $5.00

BOX.—A Practical Treatise on Heat:
As applied to the Useful Arts; for the Use of Engineers, Architects, etc. By THOMAS BOX, author of "Practical Hydraulics." Illustrated by 14 plates containing 114 figures. 12mo. $4.25

BOX.—Practical Hydraulics:
A Series of Rules and Tables for the use of Engineers, etc. By THOMAS BOX. 12mo. $2.50

BROWN.—Five Hundred and Seven Mechanical Movements:
Embracing all those which are most important in Dynamics, Hydraulics, Hydrostatics, Pneumatics, Steam Engines, Mill and other Gearing, Presses, Horology, and Miscellaneous Machinery; and including many movements never before published, and several of which have only recently come into use. By HENRY T. BROWN, Editor of the "American Artisan." In one volume, 12mo. . . . $1.00

BUCKMASTER.—The Elements of Mechanical Physics:
By J. C. BUCKMASTER, late Student in the Government School of Mines; Certified Teacher of Science by the Department of Science and Art; Examiner in Chemistry and Physics in the Royal College of Preceptors; and late Lecturer in Chemistry and Physics of the Royal Polytechnic Institute. Illustrated with numerous engravings. In one volume, 12mo. $1.50

BULLOCK.—The American Cottage Builder:
A Series of Designs, Plans, and Specifications, from $200 to $20,000, for Homes for the People; together with Warming, Ventilation, Drainage, Painting, and Landscape Gardening. By JOHN BULLOCK, Architect, Civil Engineer, Mechanician, and Editor of "The Rudiments of Architecture and Building," etc., etc. Illustrated by 75 engravings. In one volume, 8vo. $3.50

BULLOCK.—The Rudiments of Architecture and Building:
For the use of Architects, Builders, Draughtsmen, Machinists, Engineers, and Mechanics. Edited by JOHN BULLOCK, author of "The American Cottage Builder." Illustrated by 250 engravings. In one volume, 8vo. $3.50

BURGH.—Practical Illustrations of Land and Marine Engines:
Showing in detail the Modern Improvements of High and Low Pressure, Surface Condensation, and Super-heating, together with Land and Marine Boilers. By N. P. BURGH, Engineer. Illustrated by 20 plates, double elephant folio, with text. . . . $21.00

BURGH.—Practical Rules for the Proportions of Modern Engines and Boilers for Land and Marine Purposes.
By N. P. BURGH, Engineer. 12mo. $1.50

BURGH.—The Slide-Valve Practically Considered.
By N. P. BURGH, Engineer. Completely illustrated. 12mo. $2.00

BYLES.—Sophisms of Free Trade and Popular Political Economy Examined.
By a BARRISTER (Sir JOHN BARNARD BYLES, Judge of Common Pleas). First American from the Ninth English Edition, as published by the Manchester Reciprocity Association. In one volume, 12mo. Paper, 75 cts. Cloth $1.25

BYRN.—The Complete Practical Brewer:
Or Plain, Accurate, and Thorough Instructions in the Art of Brewing Beer, Ale, Porter, including the Process of making Bavarian Beer, all the Small Beers, such as Root-beer, Ginger-pop, Sarsaparilla-beer, Mead, Spruce Beer, etc., etc. Adapted to the use of Public Brewers and Private Families. By M. LA FAYETTE BYRN, M D. With illustrations. 12mo. $1.25

BYRN.—The Complete Practical Distiller:
Comprising the most perfect and exact Theoretical and Practical Description of the Art of Distillation and Rectification; including all of the most recent improvements in distilling apparatus; instructions for preparing spirits from the numerous vegetables, fruits, etc.; directions for the distillation and preparation of all kinds of brandies and other spirits, spirituous and other compounds, etc., etc. By M. LA FAYETTE BYRN, M.D. Eighth Edition. To which are added, Practical Directions for Distilling, from the French of Th. Fling, Brewer and Distiller. 12mo. $1.50

BYRNE.—Handbook for the Artisan, Mechanic, and Engineer:
Comprising the Grinding and Sharpening of Cutting Tools, Abrasive Processes, Lapidary Work, Gem and Glass Engraving, Varnishing and Lackering, Apparatus, Materials and Processes for Grinding and Polishing, etc. By OLIVER BYRNE. Illustrated by 185 wood engravings. In one volume, 8vo. $5.00

BYRNE.—Pocket Book for Railroad and Civil Engineers:
Containing New, Exact, and Concise Methods for Laying out Railroad Curves, Switches, Frog Angles, and Crossings; the Staking out of work; Levelling; the Calculation of Cuttings; Embankments; Earth-work, etc. By OLIVER BYRNE. 18mo., full bound, pocket-book form. $1.75

BYRNE.—The Practical Model Calculator:
For the Engineer, Mechanic, Manufacturer of Engine Work, Naval Architect, Miner, and Millwright. By OLIVER BYRNE. 1 volume, 8vo., nearly 600 pages $4.50

BYRNE.—The Practical Metal-Worker's Assistant:
Comprising Metallurgic Chemistry; the Arts of Working all Metals and Alloys; Forging of Iron and Steel; Hardening and Tempering; Melting and Mixing; Casting and Founding; Works in Sheet Metal; The Processes Dependent on the Ductility of the Metals; Soldering; and the most Improved Processes and Tools employed by Metal-Workers. With the Application of the Art of Electro-Metallurgy to Manufacturing Processes; collected from Original Sources, and from the Works of Holtzapffel, Bergeron, Leupold, Plumier, Napier, Scoffern, Clay, Fairbairn, and others. By OLIVER BYRNE. A new, revised, and improved edition, to which is added An Appendix, containing THE MANUFACTURE OF RUSSIAN SHEET-IRON. By JOHN PERCY, M.D., F.R.S. THE MANUFACTURE OF MALLEABLE IRON CASTINGS, and IMPROVEMENTS IN BESSEMER STEEL. By A. A. FESQUET, Chemist and Engineer. With over 600 Engravings, illustrating every Branch of the Subject. 8vo. $7.00

Cabinet Maker's Album of Furniture:
Comprising a Collection of Designs for Furniture. Illustrated by 48 Large and Beautifully Engraved Plates. In one vol., oblong $5.00

CALLINGHAM.—Sign Writing and Glass Embossing:
A Complete Practical Illustrated Manual of the Art. By JAMES CALLINGHAM. In one volume, 12mo. $1.50

CAMPIN.—A Practical Treatise on Mechanical Engineering:
Comprising Metallurgy, Moulding, Casting, Forging, Tools, Workshop Machinery, Mechanical Manipulation, Manufacture of Steam-engines, etc., etc. With an Appendix on the Analysis of Iron and Iron Ores. By FRANCIS CAMPIN, C. E. To which are added, Observations on the Construction of Steam Boilers, and Remarks upon Furnaces used for Smoke Prevention; with a Chapter on Explosions. By R. Armstrong, C. E., and John Bourne. Rules for Calculating the Change Wheels for Screws on a Turning Lathe, and for a Wheel-cutting Machine. By J. LA NICCA. Management of Steel, Including Forging, Hardening, Tempering, Annealing, Shrinking, and Expansion. And the Case-hardening of Iron. By G. EDE. 8vo. Illustrated with 29 plates and 100 wood engravings . . . $6.00

CAMPIN.—The Practice of Hand-Turning in Wood, Ivory, Shell, etc.:
With Instructions for Turning such works in Metal as may be required in the Practice of Turning Wood, Ivory, etc. Also, an Appendix on Ornamental Turning. By FRANCIS CAMPIN; with Numerous Illustrations. 12mo., cloth $3.00

CAREY.—The Works of Henry C. Carey:
FINANCIAL CRISES, their Causes and Effects. 8vo. paper . 25
HARMONY OF INTERESTS: Agricultural, Manufacturing, and Commercial. 8vo., cloth $1.50
MANUAL OF SOCIAL SCIENCE. Condensed from Carey's "Principles of Social Science." By KATE MCKEAN. 1 vol. 12mo. $2.25
MISCELLANEOUS WORKS: comprising "Harmony of Interests," "Money," "Letters to the President," "Financial Crises," "The Way to Outdo England Without Fighting Her," "Resources of the Union," "The Public Debt," "Contraction or Expansion?" "Review of the Decade 1857-'67," "Reconstruction," etc., etc. Two vols., 8vo., cloth $10.00
PAST, PRESENT, AND FUTURE. 8vo. $2.50
PRINCIPLES OF SOCIAL SCIENCE. 3 vols., 8vo., cloth $10.00
THE SLAVE-TRADE, DOMESTIC AND FOREIGN; Why it Exists, and How it may be Extinguished (1853). 8vo., cloth . $2.00
LETTERS ON INTERNATIONAL COPYRIGHT (1867) . 50
THE UNITY OF LAW: As Exhibited in the Relations of Physical, Social, Mental, and Moral Science (1872). In one volume, 8vo., pp. xxiii., 433. Cloth $3.50

CHAPMAN.—A Treatise on Ropemaking:
As Practised in private and public Rope yards, with a Description of the Manufacture, Rules, Tables of Weights, etc., adapted to the Trades, Shipping, Mining, Railways, Builders, etc. By ROBERT CHAPMAN. 24mo. $1.50

COLBURN.—The Locomotive Engine:
Including a Description of its Structure, Rules for Estimating its Capabilities, and Practical Observations on its Construction and Management. By ZERAH COLBURN. Illustrated. A new edition. 12mo. $1.25

CRAIK.—The Practical American Millwright and Miller.
By DAVID CRAIK, Millwright. Illustrated by numerous wood engravings, and two folding plates. 8vo. $5.00

DE GRAFF.—The Geometrical Stair Builders' Guide:
Being a Plain Practical System of Hand-Railing, embracing all its necessary Details, and Geometrically Illustrated by 22 Steel Engravings; together with the use of the most approved principles of Practical Geometry. By SIMON DE GRAFF, Architect. 4to. . $5.00

DE KONINCK.—DIETZ.—A Practical Manual of Chemical Analysis and Assaying:
As applied to the Manufacture of Iron from its Ores, and to Cast Iron, Wrought Iron, and Steel, as found in Commerce. By L. L. DE KONINCK, Dr. Sc., and E. DIETZ, Engineer. Edited with Notes, by ROBERT MALLET, F.R.S., F.S.G., M.I.C.E., etc. American Edition, Edited with Notes and an Appendix on Iron Ores, by A. A. FESQUET, Chemist and Engineer. One volume, 12mo. $2.50

DUNCAN.—Practical Surveyor's Guide:
Containing the necessary information to make any person, of common capacity, a finished land surveyor without the aid of a teacher. By ANDREW DUNCAN. Illustrated. 12mo., cloth. . . . $1.25

DUPLAIS.—A Treatise on the Manufacture and Distillation of Alcoholic Liquors:
Comprising Accurate and Complete Details in Regard to Alcohol from Wine, Molasses, Beets, Grain, Rice, Potatoes, Sorghum, Asphodel, Fruits, etc.; with the Distillation and Rectification of Brandy, Whiskey, Rum, Gin, Swiss Absinthe, etc., the Preparation of Aromatic Waters, Volatile Oils or Essences, Sugars, Syrups, Aromatic Tinctures, Liqueurs, Cordial Wines, Effervescing Wines, etc., the Aging of Brandy and the Improvement of Spirits, with Copious Directions and Tables for Testing and Reducing Spirituous Liquors, etc., etc. Translated and Edited from the French of MM. DUPLAIS, Ainé et Jeune. By M. MCKENNIE, M.D. To which are added the United States Internal Revenue Regulations for the Assessment and Collection of Taxes on Distilled Spirits. Illustrated by fourteen folding plates and several wood engravings. 743 pp., 8vo. $10.00

DUSSAUCE.—A General Treatise on the Manufacture of Every Description of Soap:
Comprising the Chemistry of the Art, with Remarks on Alkalies, Saponifiable Fatty Bodies, the apparatus necessary in a Soap Factory, Practical Instructions in the manufacture of the various kinds of Soap, the assay of Soaps, etc., etc. Edited from Notes of Larmé, Fontenelle, Malapayre, Dufour, and others, with large and important additions by Prof. H. DUSSAUCE, Chemist. Illustrated. In one vol., 8vo. . $10.00

DUSSAUCE.—A General Treatise on the Manufacture of Vinegar:
Theoretical and Practical. Comprising the various Methods, by the Slow and the Quick Processes, with Alcohol, Wine, Grain, Malt, Cider, Molasses, and Beets; as well as the Fabrication of Wood Vinegar, etc., etc. By Prof. H. DUSSAUCE. In one volume, 8vo. . . $5.00

DUSSAUCE.—A New and Complete Treatise on the Arts of Tanning, Currying, and Leather Dressing:
Comprising all the Discoveries and Improvements made in France, Great Britain, and the United States. Edited from Notes and Documents of Messrs. Sallerou, Grouvelle, Duval, Dessables, Labarraque, Payen, René, De Fontenelle, Malapeyre, etc., etc. By Prof. H. DUSSAUCE, Chemist. Illustrated by 212 wood engravings. 8vo. $25.00

DUSSAUCE.—A Practical Guide for the Perfumer:
Being a New Treatise on Perfumery, the most favorable to the Beauty without being injurious to the Health, comprising a Description of the substances used in Perfumery, the Formulæ of more than 1000 Preparations, such as Cosmetics, Perfumed Oils, Tooth Powders, Waters, Extracts, Tinctures, Infusions, Spirits, Vinaigres, Essential Oils, Pastels, Creams, Soaps, and many new Hygienic Products not hitherto described. Edited from Notes and Documents of Messrs. Debay, Lunel, etc. With additions by Prof. H. DUSSAUCE, Chemist. 12mo. $3.00

DUSSAUCE.—Practical Treatise on the Fabrication of Matches, Gun Cotton, and Fulminating Powders.
By Prof. H. DUSSAUCE. 12mo. $3.00

Dyer and Color-maker's Companion:
Containing upwards of 200 Receipts for making Colors, on the most approved principles, for all the various styles and fabrics now in existence; with the Scouring Process, and plain Directions for Preparing, Washing-off, and Finishing the Goods. In one vol., 12mo. . $1.25

EASTON.—A Practical Treatise on Street or Horse-power Railways.
By ALEXANDER EASTON, C. E. Illustrated by 23 plates. 8vo., cloth. $2.00

ELDER.—Questions of the Day:
Economic and Social. By Dr. WILLIAM ELDER. 8vo. . $3.00

FAIRBAIRN.—The Principles of Mechanism and Machinery of Transmission:
Comprising the Principles of Mechanism, Wheels, and Pulleys, Strength and Proportions of Shafts, Coupling of Shafts, and Engaging and Disengaging Gear. By Sir WILLIAM FAIRBAIRN, C.E., LL.D., F.R.S., F.G.S. Beautifully illustrated by over 150 wood-cuts. In one volume, 12mo. $2.50

FORSYTH.—Book of Designs for Headstones, Mural, and other Monuments:
Containing 78 Designs. By JAMES FORSYTH. With an Introduction by CHARLES BOUTELL, M. A. 4to., cloth. $5.00

GIBSON.—The American Dyer:
A Practical Treatise on the Coloring of Wool, Cotton, Yarn and Cloth, in three parts. Part First gives a descriptive account of the Dye Stuffs; if of vegetable origin, where produced, how cultivated, and how prepared for use; if chemical, their composition, specific gravities, and general adaptability, how adulterated, and how to detect the adulterations, etc. Part Second is devoted to the Coloring of Wool, giving recipes for one hundred and twenty-nine different colors or shades, and is supplied with sixty colored samples of Wool. Part Third is devoted to the Coloring of Raw Cotton or Cotton Waste, for mixing with Wool Colors in the Manufacture of all kinds of Fabrics, gives recipes for thirty-eight different colors or shades, and is supplied with twenty-four colored samples of Cotton Waste. Also, recipes for Coloring Beavers, Doeskins, and Flannels, with remarks upon Anilines, giving recipes for fifteen different colors or shades, and nine samples of Aniline Colors that will stand both the Fulling and Scouring process. Also, recipes for Aniline Colors on Cotton Thread, and recipes for Common Colors on Cotton Yarns. Embracing in all over two hundred recipes for Colors and Shades, and ninety-four samples of Colored Wool and Cotton Waste, etc. By RICHARD H. GIBSON, Practical Dyer and Chemist. In one volume, 8vo. . . $12.50

GILBART.—History and Principles of Banking:
A Practical Treatise. By JAMES W. GILBART, late Manager of the London and Westminster Bank. With additions. In one volume, 8vo., 600 pages, sheep $5.00

Gothic Album for Cabinet Makers:
Comprising a Collection of Designs for Gothic Furniture. Illustrated by 23 large and beautifully engraved plates. Oblong . . $3.00

GRANT.—Beet-root Sugar and Cultivation of the Beet.
By E. B. GRANT. 12mo. $1.25

GREGORY.—Mathematics for Practical Men:
Adapted to the Pursuits of Surveyors, Architects, Mechanics, and Civil Engineers. By OLINTHUS GREGORY. 8vo., plates, cloth $3.00

GRISWOLD.—Railroad Engineer's Pocket Companion for the Field:
Comprising Rules for Calculating Deflection Distances and Angles, Tangential Distances and Angles, and all Necessary Tables for Engineers; also the art of Levelling from Preliminary Survey to the Construction of Railroads, intended Expressly for the Young Engineer, together with Numerous Valuable Rules and Examples. By W. GRISWOLD. 12mo., tucks $1.75

GRUNER.—Studies of Blast Furnace Phenomena.
By M. L. GRUNER, President of the General Council of Mines of France, and lately Professor of Metallurgy at the Ecole des Mines. Translated, with the Author's sanction, with an Appendix, by L. D. B. Gordon, F. R. S. E., F. G. S. Illustrated. 8vo. . . . $2.50

GUETTIER.—Metallic Alloys:
Being a Practical Guide to their Chemical and Physical Properties, their Preparation, Composition, and Uses. Translated from the French of A. GUETTIER, Engineer and Director of Foundries, author of "La Fouderie en France," etc., etc. By A. A. FESQUET, Chemist and Engineer. In one volume, 12mo. $3.00

HARRIS.—Gas Superintendent's Pocket Companion.
By HARRIS & BROTHER, Gas Meter Manufacturers, 1115 and 1117 Cherry Street, Philadelphia. Full bound in pocket-book form $2.00

Hats and Felting:
A Practical Treatise on their Manufacture. By a Practical Hatter. Illustrated by Drawings of Machinery, etc. 8vo. . . . $1.25

HOFMANN.—A Practical Treatise on the Manufacture of Paper in all its Branches.
By CARL HOFMANN. Late Superintendent of paper mills in Germany and the United States; recently manager of the Public Ledger Paper Mills, near Elkton, Md. Illustrated by 110 wood engravings, and five large folding plates. In one volume, 4to., cloth; 398 pages $15.00

HUGHES.—American Miller and Millwright's Assistant.
By WM. CARTER HUGHES. A new edition. In one vol., 12mo. $1.50

HURST.—A Hand-Book for Architectural Surveyors and others engaged in Building:
Containing Formulæ useful in Designing Builder's work, Table of Weights, of the materials used in Building, Memoranda connected with Builders' work, Mensuration, the Practice of Builders' Measurement, Contracts of Labor, Valuation of Property, Summary of the Practice in Dilapidation, etc., etc. By J. F. HURST, C. E. Second edition, pocket-book form, full bound $2.50

JERVIS.—Railway Property:
A Treatise on the Construction and Management of Railways; designed to afford useful knowledge, in the popular style, to the holders of this class of property; as well as Railway Managers, Officers, and Agents. By JOHN B. JERVIS, late Chief Engineer of the Hudson River Railroad, Croton Aqueduct, etc. In one vol., 12mo., cloth $2.00

JOHNSTON.—Instructions for the Analysis of Soils, Limestones, and Manures.
By J. F. W. JOHNSTON. 12mo. 38

KEENE.—A Hand-Book of Practical Gauging:
For the Use of Beginners, to which is added, A Chapter on Distillation, describing the process in operation at the Custom House for ascertaining the strength of wines. By JAMES B. KEENE, of H. M. Customs. 8vo. $1.25

KELLEY.—Speeches, Addresses, and Letters on Industrial and Financial Questions.
By Hon. WILLIAM D. KELLEY, M. C. In one volume, 544 pages, 8vo. $3.00

KENTISH.—A Treatise on a Box of Instruments,
And the Slide Rule; with the Theory of Trigonometry and Logarithms, including Practical Geometry, Surveying, Measuring of Timber, Cask and Malt Gauging, Heights, and Distances. By THOMAS KENTISH. In one volume. 12mo. $1.25

KOBELL.—ERNI.—Mineralogy Simplified:
A short Method of Determining and Classifying Minerals, by means of simple Chemical Experiments in the Wet Way. Translated from the last German Edition of F. VON KOBELL, with an Introduction to Blow-pipe Analysis and other additions. By HENRI ERNI, M. D., late Chief Chemist, Department of Agriculture, author of "Coal Oil and Petroleum." In one volume, 12mo. $2.50

LANDRIN.—A Treatise on Steel:
Comprising its Theory, Metallurgy, Properties, Practical Working, and Use. By M. H. C. LANDRIN, Jr., Civil Engineer. Translated from the French, with Notes, by A. A. FESQUET, Chemist and Engineer. With an Appendix on the Bessemer and the Martin Processes for Manufacturing Steel, from the Report of Abram S. Hewitt, United States Commissioner to the Universal Exposition, Paris, 1867. In one volume, 12mo. $3.00

LARKIN.—The Practical Brass and Iron Founder's Guide:
A Concise Treatise on Brass Founding, Moulding, the Metals and their Alloys, etc.: to which are added Recent Improvements in the Manufacture of Iron, Steel by the Bessemer Process, etc., etc. By JAMES LARKIN, late Conductor of the Brass Foundry Department in Reany, Neafie & Co's. Penn Works, Philadelphia. Fifth edition, revised, with Extensive additions. In one volume, 12mo. . . $2.25

LEAVITT.—Facts about Peat as an Article of Fuel:
With Remarks upon its Origin and Composition, the Localities in which it is found, the Methods of Preparation and Manufacture, and the various Uses to which it is applicable; together with many other matters of Practical and Scientific Interest. To which is added a chapter on the Utilization of Coal Dust with Peat for the Production of an Excellent Fuel at Moderate Cost, specially adapted for Steam Service. By T. H. LEAVITT. Third edition. 12mo. . . . $1.75

LEROUX, C.—A Practical Treatise on the Manufacture of Worsteds and Carded Yarns:
Comprising Practical Mechanics, with Rules and Calculations applied to Spinning; Sorting, Cleaning, and Scouring Wools; the English and French methods of Combing, Drawing, and Spinning Worsteds and Manufacturing Carded Yarns. Translated from the French of CHARLES LEROUX, Mechanical Engineer, and Superintendent of a Spinning Mill, by HORATIO PAINE, M. D., and A. A. FESQUET, Chemist and Engineer. Illustrated by 12 large Plates. To which is added an Appendix, containing extracts from the Reports of the International Jury, and of the Artisans selected by the Committee appointed by the Council of the Society of Arts, London, on Woollen and Worsted Machinery and Fabrics, as exhibited in the Paris Universal Exposition, 1867. 8vo., cloth. $5.00

LESLIE (Miss).—Complete Cookery:
Directions for Cookery in its Various Branches. By MISS LESLIE. 60th thousand. Thoroughly revised, with the addition of New Receipts. In one volume, 12mo., cloth. $1.50

LESLIE (Miss).—Ladies' House Book:
A Manual of Domestic Economy. 20th revised edition. 12mo., cloth.

LESLIE (Miss).—Two Hundred Receipts in French Cookery.
Cloth, 12mo.

LIEBER.—Assayer's Guide:
Or, Practical Directions to Assayers, Miners, and Smelters, for the Tests and Assays, by Heat and by Wet Processes, for the Ores of all the principal Metals, of Gold and Silver Coins and Alloys, and of Coal, etc. By OSCAR M. LIEBER. 12mo., cloth. . . $1.25

LOTH.—The Practical Stair Builder:
A Complete Treatise on the Art of Building Stairs and Hand-Rails, Designed for Carpenters, Builders, and Stair-Builders. Illustrated with Thirty Original Plates. By C. EDWARD LOTH, Professional Stair-Builder. One large 4to. volume. $10.00

LOVE.—The Art of Dyeing, Cleaning, Scouring, and Finishing, on the Most Approved English and French Methods:
Being Practical Instructions in Dyeing Silks, Woollens, and Cottons, Feathers, Chips, Straw, etc. Scouring and Cleaning Bed and Window Curtains, Carpets, Rugs, etc. French and English Cleaning, any Color or Fabric of Silk, Satin, or Damask. By THOMAS LOVE, a Working Dyer and Scourer. Second American Edition, to which are added General Instructions for the Use of Aniline Colors. In one volume, 8vo., 343 pages. $5.00

MAIN and BROWN.—Questions on Subjects Connected with the Marine Steam-Engine:
And Examination Papers: with Hints for their Solution. By THOMAS J. MAIN, Professor of Mathematics, Royal Naval College, and THOMAS BROWN, Chief Engineer, R. N. 12mo., cloth. . . . $1.50

MAIN and BROWN.—The Indicator and Dynamometer:
With their Practical Applications to the Steam-Engine. By THOMAS J. MAIN, M. A. F. R., Assistant Professor Royal Naval College, Portsmouth, and THOMAS BROWN, Assoc. Inst. C. E., Chief Engineer, R. N., attached to the Royal Naval College. Illustrated. From the Fourth London Edition. 8vo. $1.50

MAIN and BROWN.—The Marine Steam-Engine.
By THOMAS J. MAIN, F. R.; Assistant S. Mathematical Professor at the Royal Naval College, Portsmouth, and THOMAS BROWN, Assoc. Inst. C. E., Chief Engineer R. N. Attached to the Royal Naval College. Authors of "Questions connected with the Marine Steam-Engine," and the "Indicator and Dynamometer." With numerous Illustrations. In one volume, 8vo. $5.00

MARTIN.—Screw-Cutting Tables, for the Use of Mechanical Engineers:
Showing the Proper Arrangement of Wheels for Cutting the Threads of Screws of any required Pitch; with a Table for Making the Universal Gas-Pipe Thread and Taps. By W. A. MARTIN, Engineer. 8vo. 50

Mechanics' (Amateur) Workshop:
A treatise containing plain and concise directions for the manipulation of Wood and Metals, including Casting, Forging, Brazing, Soldering, and Carpentry. By the author of the "Lathe and its Uses." Third edition. Illustrated. 8vo. $3.00

MOLESWORTH.—Pocket-Book of Useful Formulæ and Memoranda for Civil and Mechanical Engineers.
By GUILFORD L. MOLESWORTH, Member of the Institution of Civil Engineers, Chief Resident Engineer of the Ceylon Railway. Second American, from the Tenth London Edition. In one volume, full bound in pocket-book form. $2.00

NAPIER.—A System of Chemistry Applied to Dyeing.
By JAMES NAPIER, F. C. S. A New and Thoroughly Revised Edition. Completely brought up to the present state of the Science, including the Chemistry of Coal Tar Colors, by A. A. FESQUET, Chemist and Engineer. With an Appendix on Dyeing and Calico Printing, as shown at the Universal Exposition, Paris, 1867. Illustrated. In one volume, 8vo., 422 pages. $5.00

NAPIER.—Manual of Electro-Metallurgy:
Including the Application of the Art to Manufacturing Processes. By JAMES NAPIER. Fourth American, from the Fourth London edition, revised and enlarged. Illustrated by engravings. In one vol., 8vo. $2.00

NASON.—Table of Reactions for Qualitative Chemical Analysis.
By HENRY B. NASON, Professor of Chemistry in the Rensselaer Polytechnic Institute, Troy, New York. Illustrated by Colors. . 63

NEWBERY.—Gleanings from Ornamental Art of every style:
Drawn from Examples in the British, South Kensington, Indian, Crystal Palace, and other Museums, the Exhibitions of 1851 and 1862, and the best English and Foreign works. In a series of one hundred exquisitely drawn Plates, containing many hundred examples. By ROBERT NEWBERY. 4to. $15.00

NICHOLSON.—A Manual of the Art of Bookbinding:
Containing full instructions in the different Branches of Forwarding, Gilding, and Finishing. Also, the Art of Marbling Book-edges and Paper. By JAMES B. NICHOLSON. Illustrated. 12mo., cloth. $2.25

NICHOLSON.—The Carpenter's New Guide:
A Complete Book of Lines for Carpenters and Joiners. By PETER NICHOLSON. The whole carefully and thoroughly revised by H. K. DAVIS, and containing numerous new and improved and original Designs for Roofs, Domes, etc. By SAMUEL SLOAN, Architect. Illustrated by 80 plates. 4to. $4.50

NORRIS.—A Hand-book for Locomotive Engineers and Machinists:
Comprising the Proportions and Calculations for Constructing Locomotives; Manner of Setting Valves; Tables of Squares, Cubes, Areas, etc., etc. By SEPTIMUS NORRIS, Civil and Mechanical Engineer. New edition. Illustrated. 12mo., cloth. $2.00

NYSTROM.—On Technological Education, and the Construction of Ships and Screw Propellers:
For Naval and Marine Engineers. By JOHN W. NYSTROM, late Acting Chief Engineer, U. S. N. Second edition, revised with additional matter. Illustrated by seven engravings. 12mo. . . $1.50

O'NEILL.—A Dictionary of Dyeing and Calico Printing:
Containing a brief account of all the Substances and Processes in use in the Art of Dyeing and Printing Textile Fabrics; with Practical Receipts and Scientific Information. By CHARLES O'NEILL, Analytical Chemist; Fellow of the Chemical Society of London; Member of the Literary and Philosophical Society of Manchester; Author of "Chemistry of Calico Printing and Dyeing." To which is added an Essay on Coal Tar Colors and their application to Dyeing and Calico Printing. By A. A. FESQUET, Chemist and Engineer. With an Appendix on Dyeing and Calico Printing, as shown at the Universal Exposition, Paris, 1867. In one volume, 8vo., 491 pages. . $6.00

ORTON.—Underground Treasures:
How and Where to Find Them. A Key for the Ready Determination of all the Useful Minerals within the United States. By JAMES ORTON, A. M. Illustrated, 12mo. $1.50

OSBORN.—American Mines and Mining:
Theoretically and Practically Considered. By Prof. H. S. OSBORN. Illustrated by numerous engravings. 8vo. (*In preparation.*)

OSBORN.—The Metallurgy of Iron and Steel:
Theoretical and Practical in all its Branches; with special reference to American Materials and Processes. By H. S. OSBORN, LL. D., Professor of Mining and Metallurgy in Lafayette College, Easton, Pennsylvania. Illustrated by numerous large folding plates and wood-engravings. 8vo. $15.00

OVERMAN.—The Manufacture of Steel:
Containing the Practice and Principles of Working and Making Steel. A Handbook for Blacksmiths and Workers in Steel and Iron, Wagon Makers, Die Sinkers, Cutlers, and Manufacturers of Files and Hardware, of Steel and Iron, and for Men of Science and Art. By FREDERICK OVERMAN, Mining Engineer, Author of the "Manufacture of Iron," etc. A new, enlarged, and revised Edition. By A. A. FESQUET, Chemist and Engineer. $1.50

OVERMAN.—The Moulder and Founder's Pocket Guide:
A Treatise on Moulding and Founding in Green-sand, Dry-sand, Loam, and Cement; the Moulding of Machine Frames, Mill-gear, Hollowware, Ornaments, Trinkets, Bells, and Statues; Description of Moulds for Iron, Bronze, Brass, and other Metals; Plaster of Paris, Sulphur, Wax, and other articles commonly used in Casting; the Construction of Melting Furnaces, the Melting and Founding of Metals; the Composition of Alloys and their Nature. With an Appendix containing Receipts for Alloys, Bronze, Varnishes and Colors for Castings; also, Tables on the Strength and other qualities of Cast Metals. By FREDERICK OVERMAN, Mining Engineer, Author of "The Manufacture of Iron." With 42 Illustrations. 12mo. $1.50

Painter, Gilder, and Varnisher's Companion:
Containing Rules and Regulations in everything relating to the Arts of Painting, Gilding, Varnishing, Glass-Staining, Graining, Marbling, Sign-Writing, Gilding on Glass, and Coach Painting and Varnishing; Tests for the Detection of Adulterations in Oils, Colors, etc.; and a Statement of the Diseases to which Painters are peculiarly liable, with the Simplest and Best Remedies. Sixteenth Edition. Revised, with an Appendix. Containing Colors and Coloring–Theoretical and Practical. Comprising descriptions of a great variety of Additional Pigments, their Qualities and Uses, to which are added, Dryers, and Modes and Operations of Painting, etc. Together with Chevreul's Principles of Harmony and Contrast of Colors. 12mo., cloth. $1.50

HENRY CAREY BAIRD'S CATALOGUE. 17

PALLETT.—The Miller's, Millwright's, and Engineer's Guide.
By HENRY PALLETT. Illustrated. In one volume, 12mo. $3.00

PERCY.—The Manufacture of Russian Sheet-Iron.
By JOHN PERCY, M.D., F.R.S., Lecturer on Metallurgy at the Royal School of Mines, and to The Advanced Class of Artillery Officers at the Royal Artillery Institution, Woolwich; Author of "Metallurgy." With Illustrations. 8vo., paper. . . . . . . 50 cts.

PERKINS.—Gas and Ventilation.
Practical Treatise on Gas and Ventilation. With Special Relation to Illuminating, Heating, and Cooking by Gas. Including Scientific Helps to Engineer-students and others. With Illustrated Diagrams. By E. E. PERKINS. 12mo., cloth. $1.25

PERKINS and STOWE.—A New Guide to the Sheet-iron and Boiler Plate Roller:
Containing a Series of Tables showing the Weight of Slabs and Piles to produce Boiler Plates, and of the Weight of Piles and the Sizes of Bars to produce Sheet-iron; the Thickness of the Bar Gauge in decimals; the Weight per foot, and the Thickness on the Bar or Wire Gauge of the fractional parts of an inch; the Weight per sheet, and the Thickness on the Wire Gauge of Sheet-iron of various dimensions to weigh 112 lbs. per bundle; and the conversion of Short Weight into Long Weight, and Long Weight into Short. Estimated and collected by G. H. PERKINS and J. G. STOWE. $2.50

PHILLIPS and DARLINGTON.—Records of Mining and Metallurgy;
Or Facts and Memoranda for the use of the Mine Agent and Smelter. By J. ARTHUR PHILLIPS, Mining Engineer, Graduate of the Imperial School of Mines, France, etc., and JOHN DARLINGTON. Illustrated by numerous engravings. In one volume, 12mo. . . $2.00

PROTEAUX.—Practical Guide for the Manufacture of Paper and Boards.
By A. PROTEAUX, Civil Engineer, and Graduate of the School of Arts and Manufactures, and Director of Thiers' Paper Mill, Puy-de-Dôme. With additions, by L. S. LE NORMAND. Translated from the French, with Notes, by HORATIO PAINE, A. B., M.D. To which is added a Chapter on the Manufacture of Paper from Wood in the United States, by HENRY T. BROWN, of the "American Artisan." Illustrated by six plates, containing Drawings of Raw Materials, Machinery, Plans of Paper-Mills, etc., etc. 8vo. $10.00

REGNAULT.—Elements of Chemistry.
By M. V. REGNAULT. Translated from the French by T. FORREST BETTON, M.D., and edited, with Notes, by JAMES C. BOOTH, Melter and Refiner U. S. Mint, and WM. L. FABER, Metallurgist and Mining Engineer. Illustrated by nearly 700 wood engravings. Comprising nearly 1500 pages. In two volumes, 8vo., cloth. . . . $7.50

REID.—A Practical Treatise on the Manufacture of Portland Cement:
By HENRY REID, C. E. To which is added a Translation of M. A. Lipowitz's Work, describing a New Method adopted in Germany for Manufacturing that Cement, by W. F. REID. Illustrated by plates and wood engravings. 8vo. $6.00

RIFFAULT, VERGNAUD, and TOUSSAINT.—A Practical Treatise on the Manufacture of Varnishes.
By MM. RIFFAULT, VERGNAUD, and TOUSSAINT. Revised and Edited by M. F. MALEPEYRE and Dr. EMIL WINCKLER. Illustrated. In one volume, 8vo. (*In preparation.*)

RIFFAULT, VERGNAUD, and TOUSSAINT.—A Practical Treatise on the Manufacture of Colors for Painting:
Containing the best Formulæ and the Processes the Newest and in most General Use. By MM. RIFFAULT, VERGNAUD, and TOUSSAINT. Revised and Edited by M. F. MALEPEYRE and Dr. EMIL WINCKLER. Translated from the French by A. A. FESQUET, Chemist and Engineer. Illustrated by Engravings. In one volume, 650 pages, 8vo. $7.50

ROBINSON.—Explosions of Steam Boilers:
How they are Caused, and how they may be Prevented. By J. R. ROBINSON, Steam Engineer. 12mo. $1.25

ROPER.—A Catechism of High Pressure or Non-Condensing Steam-Engines:
Including the Modelling, Constructing, Running, and Management of Steam Engines and Steam Boilers. With Illustrations. By STEPHEN ROPER, Engineer. Full bound tucks . . . $2.00

ROSELEUR.—Galvanoplastic Manipulations:
A Practical Guide for the Gold and Silver Electro-plater and the Galvanoplastic Operator. Translated from the French of ALFRED ROSELEUR, Chemist, Professor of the Galvanoplastic Art, Manufacturer of Chemicals, Gold and Silver Electro-plater. By A. A. FESQUET, Chemist and Engineer. Illustrated by over 127 Engravings on wood. 8vo., 495 pages. $6.00

☞ *This Treatise is the fullest and by far the best on this subject ever published in the United States.*

SCHINZ.—Researches on the Action of the Blast Furnace.
By CHARLES SCHINZ. Translated from the German with the special permission of the Author by WILLIAM H. MAW and MORITZ MULLER. With an Appendix written by the Author expressly for this edition. Illustrated by seven plates, containing 28 figures. In one volume, 12mo. $4.25

SHAW.—Civil Architecture:
Being a Complete Theoretical and Practical System of Building, containing the Fundamental Principles of the Art. By EDWARD SHAW, Architect. To which is added a Treatise on Gothic Architecture, etc. By THOMAS W. SILLOWAY and GEORGE M. HARDING, Architects. The whole illustrated by One Hundred and Two quarto plates finely engraved on copper. Eleventh Edition. 4to., cloth. . $10.00

SHUNK.—A Practical Treatise on Railway Curves and Location, for Young Engineers.
By WILLIAM F. SHUNK, Civil Engineer. 12mo. . . $2.00

SLOAN.—American Houses:
A variety of Original Designs for Rural Buildings. Illustrated by 26 colored Engravings, with Descriptive References. By SAMUEL SLOAN, Architect, author of the "Model Architect," etc., etc. 8vo. $2.50

SMEATON.—Builder's Pocket Companion:
Containing the Elements of Building, Surveying, and Architecture; with Practical Rules and Instructions connected with the subject. By A. C. SMEATON, Civil Engineer, etc. In one volume, 12mo. $1.50

SMITH.—A Manual of Political Economy.
By E. PESHINE SMITH. A new Edition, to which is added a full Index. 12mo., cloth. $1.25

SMITH.—Parks and Pleasure Grounds:
Or Practical Notes on Country Residences, Villas, Public Parks, and Gardens. By CHARLES H. J. SMITH, Landscape Gardener and Garden Architect, etc., etc. 12mo. $2.25

SMITH.—The Dyer's Instructor:
Comprising Practical Instructions in the Art of Dyeing Silk, Cotton, Wool, and Worsted, and Woollen Goods: containing nearly 800 Receipts. To which is added a Treatise on the Art of Padding; and the Printing of Silk Warps, Skeins, and Handkerchiefs, and the various Mordants and Colors for the different styles of such work. By DAVID SMITH, Pattern Dyer. 12mo., cloth. . . . $3.00

SMITH.—The Practical Dyer's Guide:
Comprising Practical Instructions in the Dyeing of Shot Cobourgs, Silk Striped Orleans, Colored Orleans from Black Warps, Ditto from White Warps, Colored Cobourgs from White Warps, Merinos, Yarns, Woollen Cloths, etc. Containing nearly 300 Receipts, to most of which a Dyed Pattern is annexed. Also, A Treatise on the Art of Padding. By DAVID SMITH. In one volume, 8vo. Price. . . $25.00

STEWART.—The American System.
Speeches on the Tariff Question, and on Internal Improvements, principally delivered in the House of Representatives of the United States. By ANDREW STEWART, late M. C. from Pennsylvania. With a Portrait, and a Biographical Sketch. In one volume, 8vo, 407 pages. $3.00

STOKES.—Cabinet-maker's and Upholsterer's Companion:
Comprising the Rudiments and Principles of Cabinet-making and Upholstery, with Familiar Instructions, illustrated by Examples for attaining a Proficiency in the Art of Drawing, as applicable to Cabinet-work; the Processes of Veneering, Inlaying, and Buhl-work; the Art of Dyeing and Staining Wood, Bone, Tortoise Shell, etc. Directions for Lackering, Japanning, and Varnishing; to make French Polish; to prepare the Best Glues, Cements, and Compositions, and a number of Receipts particularly useful for workmen generally. By J. STOKES. In one volume, 12mo. With Illustrations. . $1.25

Strength and other Properties of Metals:
Reports of Experiments on the Strength and other Properties of Metals for Cannon. With a Description of the Machines for testing Metals, and of the Classification of Cannon in service. By Officers of the Ordnance Department U. S. Army. By authority of the Secretary of War. Illustrated by 25 large steel plates. In one volume, 4to. . $10.00

SULLIVAN.—Protection to Native Industry.
By Sir EDWARD SULLIVAN, Baronet, author of "Ten Chapters on Social Reforms." In one volume, 8vo. $1.50

Tables Showing the Weight of Round, Square, and Flat Bar Iron, Steel, etc.,
By Measurement. Cloth. 63

TAYLOR.—Statistics of Coal:
Including Mineral Bituminous Substances employed in Arts and Manufactures; with their Geographical, Geological, and Commercial Distribution and Amount of Production and Consumption on the American Continent. With Incidental Statistics of the Iron Manufacture. By R. C. TAYLOR. Second edition, revised by S. S. HALDEMAN. Illustrated by five Maps and many wood engravings. 8vo., cloth. $10.00

TEMPLETON.—The Practical Examiner on Steam and the Steam-Engine:
With Instructive References relative thereto, arranged for the Use of Engineers, Students, and others. By WM. TEMPLETON, Engineer. 12mo. $1.25

THOMAS.—The Modern Practice of Photography.
By R. W. THOMAS, F. C. S. 8vo., cloth. 75

THOMSON.—Freight Charges Calculator.
By ANDREW THOMSON, Freight Agent. 24mo. . . . $1.25

TURNING: Specimens of Fancy Turning Executed on the Hand or Foot Lathe:
With Geometric, Oval, and Eccentric Chucks, and Elliptical Cutting Frame. By an Amateur. Illustrated by 30 exquisite Photographs. 4to. $3.00

Turner's (The) Companion:
Containing Instructions in Concentric, Elliptic, and Eccentric Turning: also various Plates of Chucks, Tools, and Instruments; and Directions for using the Eccentric Cutter, Drill, Vertical Cutter, and Circular Rest; with Patterns and Instructions for working them. A new edition in one volume, 12mo. $1.50

URBIN.—BRULL.—A Practical Guide for Puddling Iron and Steel.
By ED. URBIN, Engineer of Arts and Manufactures. A Prize Essay read before the Association of Engineers, Graduate of the School of Mines, of Liege, Belgium, at the Meeting of 1865-6. To which is added A COMPARISON OF THE RESISTING PROPERTIES OF IRON AND STEEL. By A. BRULL. Translated from the French by A. A. FESQUET, Chemist and Engineer. In one volume, 8vo. $1.00

VAILE.—Galvanized Iron Cornice-Worker's Manual:
Containing Instructions in Laying out the Different Mitres, and Making Patterns for all kinds of Plain and Circular Work. Also, Tables of Weights, Areas and Circumferences of Circles, and other Matter calculated to Benefit the Trade. By CHARLES A. VAILE, Superintendent "Richmond Cornice Works," Richmond, Indiana. Illustrated by 21 Plates. In one volume, 4to. $5.00

VILLE.—The School of Chemical Manures:
Or, Elementary Principles in the Use of Fertilizing Agents. From the French of M. GEORGE VILLE, by A. A. FESQUET, Chemist and Engineer. With Illustrations. In one volume, 12 mo. . . $1.25

VOGDES.—The Architect's and Builder's Pocket Companion and Price Book:
Consisting of a Short but Comprehensive Epitome of Decimals, Duodecimals, Geometry and Mensuration; with Tables of U. S. Measures, Sizes, Weights, Strengths, etc., of Iron, Wood, Stone, and various other Materials, Quantities of Materials in Given Sizes, and Dimensions of Wood, Brick, and Stone; and a full and complete Bill of Prices for Carpenter's Work; also, Rules for Computing and Valuing Brick and Brick Work, Stone Work, Painting, Plastering, etc. By FRANK W. VOGDES, Architect. Illustrated. Full bound in pocketbook form. $2.00
Bound in cloth. 1.50

WARN.—The Sheet-Metal Worker's Instructor:
For Zinc, Sheet-Iron, Copper, and Tin-Plate Workers, etc. Containing a selection of Geometrical Problems; also, Practical and Simple Rules for describing the various Patterns required in the different branches of the above Trades. By REUBEN H. WARN, Practical Tinplate Worker. To which is added an Appendix, containing Instructions for Boiler Making, Mensuration of Surfaces and Solids, Rules for Calculating the Weights of different Figures of Iron and Steel, Tables of the Weights of Iron, Steel, etc. Illustrated by 32 Plates and 37 Wood Engravings. 8vo. $3.00

WARNER.—New Theorems, Tables, and Diagrams for the Computation of Earth-Work:
Designed for the use of Engineers in Preliminary and Final Estimates, of Students in Engineering, and of Contractors and other non-professional Computers. In Two Parts, with an Appendix. Part I.—A Practical Treatise; Part II.—A Theoretical Treatise; and the Appendix. Containing Notes to the Rules and Examples of Part I.; Explanations of the Construction of Scales, Tables, and Diagrams, and a Treatise upon Equivalent Square Bases and Equivalent Level Heights. The whole illustrated by numerous original Engravings, comprising Explanatory Cuts for Definitions and Problems, Stereometric Scales and Diagrams, and a Series of Lithographic Drawings from Models, showing all the Combinations of Solid Forms which occur in Railroad Excavations and Embankments. By JOHN WARNER, A. M., Mining and Mechanical Engineer. 8vo. $5.00

WATSON.—A Manual of the Hand-Lathe:
Comprising Concise Directions for working Metals of all kinds, Ivory, Bone and Precious Woods; Dyeing, Coloring, and French Polishing; Inlaying by Veneers, and various methods practised to produce Elaborate work with Dispatch, and at Small Expense. By EGBERT P. WATSON, late of "The Scientific American," Author of "The Modern Practice of American Machinists and Engineers." Illustrated by 78 Engravings. $1.50

WATSON.—The Modern Practice of American Machinists and Engineers:
Including the Construction, Application, and Use of Drills, Lathe Tools, Cutters for Boring Cylinders, and Hollow Work Generally, with the most Economical Speed for the same; the Results verified by Actual Practice at the Lathe, the Vice, and on the Floor. Together with Workshop Management, Economy of Manufacture, the Steam-Engine, Boilers, Gears, Belting, etc., etc. By EGBERT P. WATSON, late of the "Scientific American." Illustrated by 86 Engravings. In one volume, 12mo. $2.50

WATSON.—The Theory and Practice of the Art of Weaving by Hand and Power:
With Calculations and Tables for the use of those connected with the Trade. By JOHN WATSON, Manufacturer and Practical Machine Maker. Illustrated by large Drawings of the best Power Looms. 8vo. $10.00

WEATHERLY.—Treatise on the Art of Boiling Sugar, Crystallizing, Lozenge-making, Comfits, Gum Goods.
12mo. $2.00

WEDDING.—The Metallurgy of Iron;
Theoretically and Practically Considered. By Dr. HERMANN WEDDING, Professor of the Metallurgy of Iron at the Royal Mining Academy, Berlin. Translated by JULIUS DU MONT, Bethlehem, Pa. Illustrated by 207 Engravings on Wood, and three Plates. In one volume, 8vo. (*In press.*)

WILL.—Tables for Qualitative Chemical Analysis.
By Professor HEINRICH WILL, of Giessen, Germany. Seventh edition. Translated by CHARLES F. HIMES, Ph. D., Professor of Natural Science, Dickinson College, Carlisle, Pa. . . . $1.50

WILLIAMS.—On Heat and Steam:
Embracing New Views of Vaporization, Condensation, and Explosions. By CHARLES WYE WILLIAMS, A. I. C. E. Illustrated. 8vo. $3.50

WOHLER.—A Hand-Book of Mineral Analysis.
By F. WOHLER, Professor of Chemistry in the University of Göttingen. Edited by HENRY B. NASON, Professor of Chemistry in the Rensselaer Polytechnic Institute, Troy, New York. Illustrated. In one volume, 12mo. $3 00

WORSSAM.—On Mechanical Saws:
From the Transactions of the Society of Engineers, 1869. By S. W. WORSSAM, Jr. Illustrated by 18 large plates. 8vo. . . $5.00

www.ingramcontent.com/pod-product-compliance
Lightning Source LLC
Chambersburg PA
CBHW032001230426
43672CB00010B/2231